ALSO BY JERRY DENNIS

Canoeing Michigan Rivers

It's Raining Frogs and Fishes

A Place on the Water

The Bird in the Waterfall

The River Home

From a Wooden Canoe

Leelanau: A Portrait of Place in Photographs and Text
(with photos by Ken Scott)

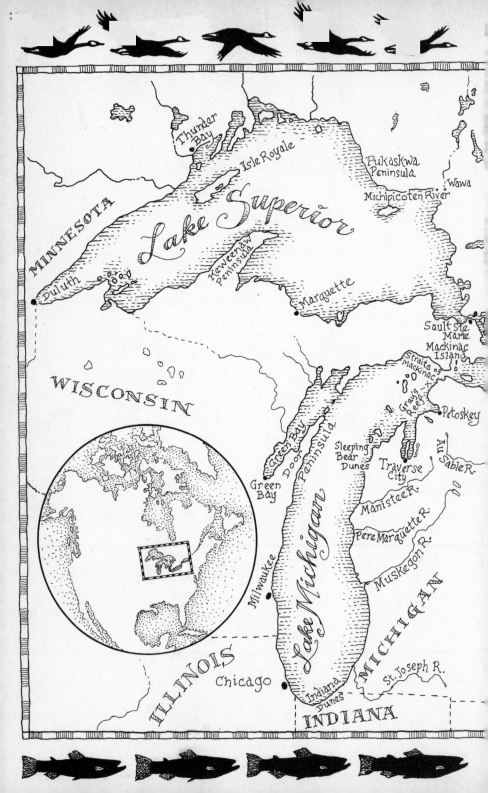

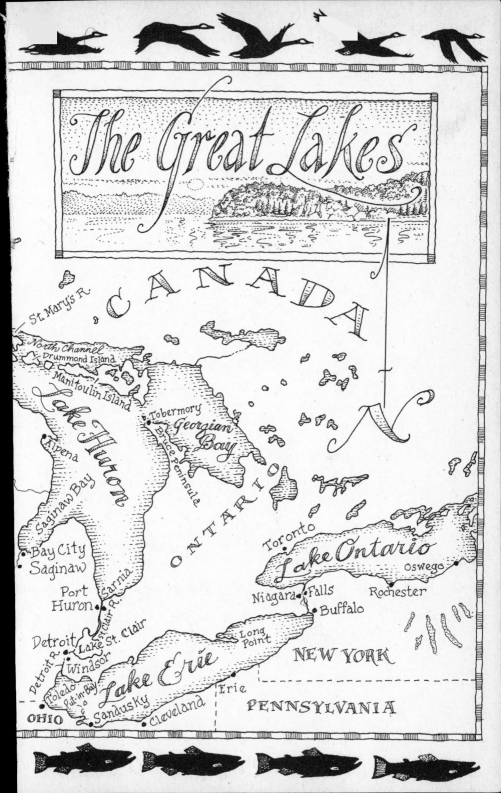

THE LIVING GREAT LAKES

Searching for the Heart of the Inland Seas

JERRY DENNIS

Thomas Dunne Books
St. Martin's Griffin ❧ *New York*

For my sister, Melissa,
with love and
admiration

THOMAS DUNNE BOOKS.
An imprint of St. Martin's Press.

www.stmartins.com

Photographs by Jerry Dennis
Map by Glenn Wolff

The author wishes to thank the editors of publications in which portions of this book have
appeared: A portion of chapter 6 was first published in *Canoe Journal*. A portion of chapter 10
originally appeared in *Sports Afield*. Chapter 12, Lake Michigan [Lake Squall, 1967], was
included in somewhat different form in the anthology *Soul of the Sky: Exploring the Human
Side of Weather* (North Conway, NH: The Mount Washington Observatory, 1999).

Library of Congress Cataloging-in-Publication Data

Dennis, Jerry.
    The living Great Lakes : searching for the heart of the inland seas / Jerry Dennis.—1st ed.
        p.   cm.
    Includes index (page 289).
    ISBN 0-312-25193-9 (hc)
    ISBN 0-312-33103-7 (pbk)
    EAN 978-0312-33103-0
    1. Great Lakes—History. 2. Natural history—Great Lakes. 3. Dennis, Jerry—
Journeys—Great Lakes Region. 4. Great Lakes—Environmental conditions. I. Title.

F551 .D39   2003
977—dc21

                                                                2002032500

First St. Martin's Griffin Edition: June 2004
20 19 18 17 .16 15 14 13 12

Contents

LAKE MICHIGAN

To appreciate the magnitude of the Great Lakes you must get close to them. Launch a boat on their waters or hike their beaches or climb the dunes, bluffs, and rocky promontories that surround them and you will see, as people have seen since the age of glaciers, that these lakes are pretty damned big. It's no wonder they're sometimes upgraded to "Inland Seas" and "Sweetwater Seas." Calling them lakes is like calling the Rockies hills. Nobody pretends they compare to the Atlantic or Pacific, but even the saltiest saltwater mariners have been surprised to discover that the lakes contain a portion of ocean fury.

The first time I saw Lake Michigan, I thought it must be an ocean. I was five years old, and my family had just moved to the Leelanau Peninsula, the little finger of Michigan's mitten, and rented a hilltop house with a view of the lake. In the living room, centered before the picture window, was a brass telescope mounted on a pedestal, where I would stand on a chair at night and peer at ships on the horizon, each lit as brightly as a small city. My father told me that they were ships five-hundred to a thousand-feet long, with cargo holds that could carry a hundred trainloads of wheat or iron ore. If they were headed south, they were probably bound for Chicago; if north, for Detroit, New York, London, Hong Kong. I would stand in our house and watch those large, bright, slowly passing vessels and sense connection with the world.

It was a magical place to live. Our yard ran in a long slope down to the lilypads of South Bar Lake, with Lake Michigan a stone's throw beyond. At the big lake was a beach empty of people most days and a playground of sandblasted swings and teeter-totters set precariously a few feet above storm waves. My memories of that summer are filled with painted turtles and watersnakes, with excursions down the beach in search of treasures, with ominous dark thunderstorms passing over the lake, lightning flashing in the distance. My mother had grown up a few miles down the shore in Glen Arbor, and my father's parents owned a cherry farm and sugarbush a few miles inland, so for them it was a homecoming. For me it was a revelation.

I remember standing in the sand, feeling very small. Gulls kited stationary above me, then banked their wings against the wind and soared away. The wind was cool and fresh and smelled like rain. A wave curled and broke; water rushed up on the sand, spread thin, and sank. The shore stretched for as far as I could see, from the haze-obscured curve of Platte Bay to the massive yellow flank of Sleeping Bear Dunes. The lake was too vast for comprehension. It was nothing but water to the edge of the earth. I thought sharks swam out there, and pirate ships sailed, and on its far shores lived people who spoke strange languages. I assumed the water was salty.

That summer my mother led my brother and me up a trail to the summit of the Empire Bluffs, a mountain of sand shoved up thousands of years ago by glaciers. It was grown over with stunted trees and dune grasses and capped with the long-dead trunks of ancient cedars bleached pale by time and weather. The view from the top was stunning. Down the shore was a strip of yellow beach between the lake and the bunched-up hills of forest, with the dunes looming beyond.

The bluff beneath us was so steep it was disorienting. I threw a stone thinking it would soar to the lake, but it struck sand a ridiculously short distance below me. I looked down at the water near shore and saw three black fish as big as logs patrolling in the shallows. Sturgeon, I now realize— the largest inhabitants of the Great Lakes and rarely encountered, though a century ago they were so abundant that farmers around the lakes pitch-forked them during their spawning runs and used them for fertilizer. The image of the gigantic fish had mythic weight. For years I wondered if I had dreamed it.

Now, forty years later, the Empire Bluffs are sheltered within Sleeping Bear Dunes National Lakeshore, and are visited more frequently than when I was a child. Not much else has changed. The Bluffs are still topped with a ghost forest of cedars, lake and sky merge seamlessly at the horizon, Sleeping Bear Dunes tip to the water like a large golden pyramid. No sturgeon swam into view the last time I visited, but I expected none—not many remain in the lake. Down the shore, the beach in Empire was crowded with people, but I expected that also. As I looked over the water, a British tourist in shorts and hiking boots climbed huffing behind me and asked in a Piccadilly lilt, "Can one see Wisconsin?"

No, sorry, one cannot. Not even with the strongest telescope. Cross Lake Michigan by boat—cross any of the Great Lakes—and most of the way there's nothing to see but water and sky. Here if you head west the crossing is roughly sixty miles of open lake to Wisconsin's Door Peninsula, with Green Bay beyond it. Green Bay, incidentally, is where the French trader Jean Nicolet, who was probably the first European to enter Lake Michigan, went ashore in a canoe in 1634 firing pistols in the air and wearing a silk robe embroidered with flowers and birds. He thought he had reached China. When no representatives of the Khan showed up to welcome him, he marched into a nearby Winnebago village and repeated his performance, no doubt providing much entertainment for the locals.

Those of us who live near the lakes take their great size for granted. We also take for granted that travel in the region is made inconvenient by water, and that in winter we're likely to be buried in "lake-effect" snow when cold, dry, Arctic winds pick up moisture and heat as they pass over the lakes, conjuring ten, twenty, and in a few places as much as thirty feet of snow a year along our coastal snowbelts. We learn in elementary school that the acronym HOMES is a handy way to remember the names of the lakes. We're taught that their surface area of 94,676 square miles is roughly the size of New York, New Jersey, Connecticut, Rhode Island, Massachusetts, New Hampshire, and Vermont combined (and is slightly larger than England, Wales, Scotland, and Northern Ireland); that the shorelines of the five lakes extend for more than 10,000 miles, about equal to the combined Atlantic and Pacific coasts of the United States; and that Michigan alone is bounded by 3,200 miles of coastline—only Alaska has more.

Give us an opportunity and we'll remind you that the lakes contain nearly a fifth of the freshwater on the surface of the planet; that if it were possible to pour all the water from all the ponds, lakes, rivers, and reservoirs in the United States into a hundred gigantic buckets, ninety-five of them would be filled by the Great Lakes; that if you distributed the water from those ninety-five buckets evenly across the land, it would cover the lower forty-eight states in a lake ten feet deep.

We can be fiercely protective, as politicians have learned, sometimes to their dismay. When Texas congressman Dick Armey came to Michigan a few years ago to endorse a local Republican candidate for Congress, he looked at Lake Michigan and said he knew a few ranchers back home who'd like to poke a siphon in *that*. Cribbing clumsily from Mark Twain, he said, "I'm from Texas and down there we understand that the whiskey is for drinking and the water is for fighting over." His point was that if we were to give up local control, Washington bureaucrats would be sure to take charge of the water. "If we get it in Washington," Armey said, speaking for thirsty Texans, "we're not going to be buying it. We'll be stealing it. You are going to have to protect your Great Lakes." By "protect" he meant, of course, defend our right to profit from it. But Great Lakes water is not Texas crude, and it's not for sale. His candidate lost.

The lakes extend 575 miles from the north shore of Lake Superior to the south shore of Lake Erie, a spread of eight degrees in latitude. From west to east they stretch nearly eight hundred miles. Their drainage basin encompasses 200,000 square miles, an area almost as big as France. In that basin live thirty-four million people, each of them affected in ways large and small by the lakes.

Anywhere you go in the region, the vernacular designates the nearest Great Lake as "the Big Lake." Each Big Lake is different, with its own character and characteristics, but the same water flows through them and they share many qualities. All five shape the land and alter the weather and define the journeys of those who live nearby.

Circumnavigation is an ambitious undertaking. From where I live, in the northwest corner of Michigan's Lower Peninsula, I can fly in a commuter plane across Lake Michigan and be in Milwaukee in less than an hour. Or, if it's May through October, I can book passage on a car ferry, the SS *Badger*, and cross from Ludington to Manitowoc in four hours. If I choose to drive to Milwaukee, I can go south around the bottom of the lake, through Chicago, then up the coast of Wisconsin; or go north around the top of the lake and down Wisconsin. Either way, the trip is four hundred miles. That's half the distance around just one lake, and not the largest.

Drive U.S. highways from the eastern end of Lake Ontario to the western end of Lake Superior and you pass through upstate New York, a corner of Pennsylvania, most of the length of Ohio, sections of Indiana and Illinois, a good share of Wisconsin, and the slanted northeast border of Minnesota; take the shortcut through Michigan and you have to drive the length of both peninsulas. Returning by the northern route, it's all Ontario. From north to south and west to east you pass through distinct ecological zones, from boreal forests to hardwood forests to till plains to clay plains to corn-belt plains to lake plains—forests in the north, farms and industry in the south, with abandoned cutovers and vestigial prairies and abundant wetlands throughout.

Alexis de Tocqueville sailed on the Great Lakes in 1831, when he was twenty-six years old, during the tour of the United States that inspired his book, *Democracy in America*. While aboard the steamboat *Ohio*, bound for Detroit along the south shore of Lake Erie, he wrote in a letter: "This lake without sails, this shore which does not yet show any trace of the passage of man, this eternal forest which borders it; all that, I assure you, is not grand in poetry only; it's the most extraordinary spectacle that I have seen in my life."

Young Tocqueville and his companion, Gustave Beaumont, crossed Erie to Detroit, coasted the west shore of Lake Huron and steamed up the St. Mary's River to the village of Sault Ste. Marie, where they glimpsed Superior ("This lake much resembles all the others," wrote Beaumont, but he was mistaken), then returned to Huron, passed through the Straits of Mackinac, and crossed Lake Michigan to Green Bay. They witnessed wonders, but they missed more than they saw. They did not see Lake Ontario, with its wooded shores and cobblestone beaches, the clay faces of the Scarborough Bluffs, the thousand islands at the lake's outlet to the St. Lawrence. They missed Lake Huron's Bruce Peninsula, with its limestone cliffs tumbling to the water, and enormous Georgian Bay and its thousands of clustered islands, its fjords, and its water-sculpted escarpments. They missed the sand mountains along the east shore of Lake Michigan, and the young city of Chicago, which in 1831 was on the verge of booming. Especially they missed Superior. They never saw the mountain range called the Sleeping Giant, its peaks a clear profile of feet, knees, belly, chest, and face. They missed the mineral-stained Pictured Rocks, storm-battered and brilliant with colors; the wild and lofty Palisade Head Cliffs; the pine-covered Porcupine Mountains, which though only two thousand feet above sea level were once as high as the Rockies. They missed most of the wildlife—the Atlantic salmon that ran up Lake Ontario's rivers to spawn, the whitefish and lake trout and sturgeon in all the lakes, the deer and moose, the bear and wolf, the flocks of passenger pigeons that migrated in numbers so vast they blackened the sky for days as they passed, yet would be gone forever by the end of the century. And they missed the changes of the seasons— never saw maples turn scarlet in October or trilliums fill the woods with blossoms in May, did not witness snow squalls racing across the lakes or surf exploding against mountains of ice along shore.

Though I've lived near the Great Lakes most of my life, there came a day a few years ago when I realized how little I knew of them. To get better acquainted, I drove around each of their shores. Eventually I drove around them again. I explored beaches and shoreline villages and city lakefronts. I met passionate people who showed me the places they loved and were fighting to protect. I filled boxes with brochures, pamphlets, reports, books. I took notes and photographs. In the end I got to know some of the people, cities, and roads—but not the lakes.

For a month and a half I stayed alone in a house on the north shore of

Lake Michigan. Mornings I worked at a desk in front of a sliding-glass door with a view of North Manitou Island, low and darkly wooded, and beyond it the horizon of the open lake blurring with the sky. Afternoons I walked the beach. It was February and March of an unusually warm winter, and I had the shore to myself. I would follow a trail from the house to the foredunes, walking through snow in February, then, in March, after the snow melted, on sand. Pausing at the bluff, I would look up and down the length of the bay. A few miles to the north was Whaleback, a wooded promontory in outline shaped like a giant sperm whale—Moby-Dick beached and grown over with forest, his blunt head yearning lakeward, his fluke raised behind. To the south, beyond the long swerve of the bay, was Pyramid Point, a raw sandy dune topped with forest. From a distance the Point looks like someone once tipped a knife at an angle and carved it smooth.

Every afternoon I walked along the same stretch of uninhabited beach and watched the ways it changed. I became interested in the relationship between sand and wind. High on the beach, where the sand was dry, was a lunar landscape I had never noticed in my years of exploring Lake Michigan beaches. Scattered across it were thousands of stones the size of golf-balls, each stranded on a pedestal of sand and casting a thin shadow. I learned that geologists call the stones "lag gravel," and that they are stranded there when wind blows the sand away from them. Larger stones that stay in place for years become faceted on the side facing the prevailing wind. Geologists call them "ventifacts."

I became interested also in the kinds of waves I saw. The smallest were capillary waves, hardly more than wrinkles on the surface of the water, which act like tiny sails to catch the wind and make larger waves. Gusts blowing over the land plummeted to the water, flurried into cat's-paws, then gathered force and raced away toward Wisconsin. Whitecaps marched across the bay and pumped up and down against the horizon line, their tips bright as snow against the blue of the lake. Breakers purled and galloped down the shore. Low swells made sluggish by the cold seemed to rise from the bottom of the lake and crawl to shore, finally collapsing on the sand like exhausted swimmers.

From Walter J. Hoagman's genial little guidebook, *Great Lakes Coastal Plants*, I learned the parts of the coastal zone. The fringe where the sand is always wet is called the "swash zone." The dry beach, above the reach of ordinary waves, is the "backshore." "Bluffs" are banks built over millennia, rising a few feet to a few hundred feet above the backshore. "Foredunes" are uneven, hilly dunes, well above the high-water mark, scattered with coastal plants. "Backdunes" are larger hills of sand, where trees and shrubs live among coastal flowers and grasses, punctuated by "blowouts" of barren sand, eroded by wind.

In the foredunes and backdunes I examined winter weeds, trying to identify by stalks and dried leaves such plants as sand cress and sandwort, fringed gentian, yarrow, false heather, and silverweed. After a few weeks I was as enchanted with the names as I was with the plants they designated. Guidebook in hand I walked the beach, reciting into the wind:

> *Lake tansy, calamint, Queen Anne's lace.*
> *Little bluestem and horsemint.*
> *Mossy stonecrop.*
> *Starry false Solomon's Seal.*
> *Sea rocket and beach pea.*
> *Soapberry, pigweed, and spiked lobelia.*
> *Indian paintbrush.*
> *Seaside spurge.*
> *Bugleweed, horsetail, windflower.*

Six weeks on the beach, and I never got tired of it. On the contrary—I wanted more. I wanted to see it all and know everything about it. Gradually I began to know those two miles of beach and dunes. But of course it wasn't the same as knowing the lake.

The following summer I stood on a ledge looking into the deep, clear water of Lake Superior. I was on its largest island, Isle Royale (pronounce it "I'll Royal"—"Eel roy-AL" brands you an outsider and a fancy-pants). It's a big place, ten times the size of Manhattan, and raw with rock and bog and impenetrable spruce forest. It is among the least visited of our national parks, but it can't bear much use, and the few hundred visitors who come each day in the summer are probably too many. The island is home to moose and wolves—their dynamic here is among the most carefully studied in the history of wildlife biology—and is dotted with inland lakes and long protected finger-bays of Superior where loons warble and moose wander down in the evenings to drink. My wife and I had come to canoe, hike, and camp. We never wanted to leave.

One day Gail and I walked a portage trail with three young biologists who were on the island studying loons. They were bright-eyed and tanned and wind-burned, glowing with that radiance you encounter now and then in people who are doing exactly what they were put on earth to do. They told us in detail about their work, about banding loons and following them from bay to bay around the island, keeping a careful distance while observing them through spotting scopes mounted on their kayaks. They'd been tracking the same birds for three years. Curious to see their reaction, I asked, "Honestly, don't you ever get a little tired of loons?" and they looked at me with their mouths hanging open. Finally one said the words that all three were thinking: "Are you crazy?"

Maybe. I'd been tracking the Great Lakes for three years by now and was beginning to think the task was hopeless. I'd become lost in the parts. Wherever I went, I wanted to know the water and everything in it and near it. I wanted to know the rocks around the shore, the insects that lived among the rocks, the birds that fed on the insects and nested in the trees, the trees themselves. And not just their names. Their life histories, their places in the whole, the poetry, philosophy, and science they had inspired in people like the loon researchers, who had devoted their lives to them. And I wanted the words to put it all together—every place, every moment, and all they signified.

It had become overwhelming. The water alone was defeating me. How do you describe water? What words can evoke those spangles of sunlight, those shifting wave shadows, those pellucid blue depths? I lacked the vocabulary. I wanted to take hold of the immediate world, see it independent of the names we give it, then give it name. But I couldn't grasp it. People five thousand years ago rode these same waters in canoes, then painted rocks with images of what they saw. I suspect that they too were unable to grasp the whole.

Emerson said the world lacks unity, or seems to, only if we have lost unity within ourselves. He thought a naturalist might learn to see the world whole, but only if all the demands of his spirit were met. "Love," he wrote, "is as much its demand as perception."

I had the love, I think, but not the perception. I couldn't see far enough. And I couldn't unite what I saw with what I already knew. I stood on that ledge above Lake Superior and looked down through the water at rocks the size of houses, but I couldn't get to them. I couldn't get to anything. Before me was water, billions of Mickey Mouse molecules in every drop, and every drop as pristine as mountain air, flavored with cedar and feldspar, colored with sky, granite, and spruce. I didn't want to trivialize what I saw, and to dissect it would murder it. I'd done enough dissecting. I was reaching for something else entirely. I wanted to hold what I saw, felt, heard, tasted, and scented, and to possess it always—not like a tourist snapping photos, but literally, taking possession of its physical fact and keeping it with me always—yet I couldn't get my arms around it.

It occurred to me that I should strip off my shirt, raise myself on my toes, breathe deeply, and dive. Immerse myself. Swim down into emerald depths until the weight of the lake embraced me and I could run my hands over granite blocks that had never been touched. It would have been still and cool down there, and very quiet.

But I lacked courage. The water was too cold by far. I thought the shock might burst my heart.

So I stood safe and dry on shore and looked across all those miles of Lake Superior and saw all that I was missing—and decided I needed a boat.

Chapter 2

LAKE MICHIGAN

A Meeting on the *Malabar* ◆ The History of a Concrete Boat ◆
A Providential Seiche ◆ We Begin Our Voyage

Hajo Knuttel knew the lakes were big, but not *that* big. Hajo—pronounced HI-oh, "like Ohio, without the first O"—was a fifty-year-old freelance ship's captain from Connecticut who had spent most of his life sailing the oceans but was seeing the Great Lakes for the first time. He was surprised at the blueness of the water, which he compared to the Caribbean, at its apparent cleanliness, at its magnitude. "Pretty big ponds," he would say a dozen times during the weeks I sailed with him across four of the lakes (Huron, Ontario, Michigan, and Erie—until he visits Superior, HOME is all he knows). He would be deeply curious about the human and geological histories of the lakes, about the plants and animals that live in and around them, about their environmental and biological degradations.

Hajo and I met one cold and gusty afternoon in April 2000 at Harbor West Marina on Grand Traverse Bay, near Traverse City, Michigan, where I live. I had walked the length of the marina dock, past pleasure cruisers and private sailing yachts, to a handsome, tall-masted schooner teed at the end. At more than a hundred feet in length, with a pair of sixty-foot masts, and displacing seventy-four tons, the *Malabar* dwarfed every boat in the harbor.

I was no sailor, but I knew enough to seek permission before stepping aboard. The deck was strewn with 55-gallon drums and stacked with cans of paint and coils of rigging, but nobody was in sight. I stood on the dock for a moment and finally called out, "Anybody home?"—a landlubber's expression for sure. A man popped his head through a hatch in the deck and appraised me.

He had blue eyes and longish blond-gray hair parted boyishly off center. On his cheeks grew muttonchops of the sort you'd expect to see on eighteenth-century sea captains. His expression was curious but guarded.

"Hello," I said. "Are you Hajo?"

"Who are you?"

I explained that the owner of the boat had suggested I talk to him. I knew he was sailing the *Malabar* through the lakes and ultimately to Maine, and I was hoping to catch a ride part of the way. I was writing a book and thought a schooner might offer an interesting view of the Great Lakes. With his permission I would like to go as far as the St. Lawrence River, then jump ship and catch a plane back to Michigan.

"You willing to work?" he asked.

"Of course."

"Stand night watch?"

"Yes."

"Because this isn't a pleasure cruise. There's a lot of work to be done, and everyone has to do his share."

"I understand."

"Come aboard, then."

He led the way down a ladder into a large and cluttered cabin paneled all around with knotty pine. "Sit down, sit down," he said, pointing to a bench at a table. A doorway led to a darkened room, where I could see a bunk heaped with clothing—the captain's cabin. Another doorway opened into a galley dominated by a stainless-steel stove. Woodwork everywhere was varnished to a gleam. One wall was covered with dials and gauges. Another was pinned with a large road map of the eastern United States, the Great Lakes prominent. All around the cabin were recessed berths, enough to sleep a dozen or more, each stuffed with tools, boxes of groceries, stacks of orange life jackets. A pair of tables filled most of the floor space. They were covered with books, charts, manuals, wrenches, screwdrivers.

"You're writing a book? About what?"

The Great Lakes, I said, a sort of biography. It was hard to explain. He asked me to try, so I told him a little about my research so far.

He admitted that all he'd seen of the lakes was a bit of Erie from the window of his jet as he flew to Detroit, and a glimpse of Lake Michigan as he drove into Traverse City. Most of the view was obscured by a snowstorm.

"They said it was something called lake-effect snow. What's that?"

I explained it as best I could. For it to occur, a rare set of circumstances must be met. A body of water has to be located at a latitude cold enough to produce snow—but not so cold that the water freezes over—and must be large enough to warm the air that passes over it—but not warm it much above freezing. Also required is a mass of land that supplies cold air upwind of the water. Such conditions are met only in the Great Lakes and along a few ocean coasts—the east shore of Hudson Bay, for instance, and the west coast of the Japanese islands of Honshu and Hokkaido. Generally, the greater the temperature difference between the water and the air, the greater the snowfall. It also varies with fetch, or the distance the wind travels over open water. Places downwind of the longest fetches—the south shore of Lake Superior and the east and south shores of the other lakes—receive legendary snows. Buffalo gets buried. So does Rochester. The most snow on record is on the Keweenaw Peninsula, at the Superior shore of Michigan's Upper Peninsula, where more than thirty-two feet fell in the winter of 1978–79.

Hajo considered this information. He was a careful listener. In the month and a half that he had been living aboard the *Malabar* he'd become deeply interested in his surroundings. He was intrigued, for example, by the clarity of the water in the harbor. He could look over the side of the boat and watch trout and pike swim past on the bottom. It surprised him. He'd heard the Great Lakes were polluted.

The lakes were clearer now than at any time in our lives, I said, partly as a result of more stringent pollution controls, but also because of the zebra mussel, an invader from Eastern Europe that had spread throughout the lakes and was filtering organisms from the water. This, too, interested Hajo. It turned out that he had a bachelor's degree in marine biology, read widely on the subject, and could rattle off both the common and Latin names of many plants, fishes, birds, and microorganisms of the sea. But freshwater organisms were less familiar to him. He seemed ready to roll up his sleeves and begin learning.

Suddenly Hajo lunged into the captain's cabin. He returned with a book in his hand. "Have you read this?" he asked, and thrust it at me. Before I could answer, he dove into the cabin again and returned with two other books. He pressed them upon me, extolling enthusiastically. "You have to understand," he said. "I would die without books." I glanced at them: Villiers, Forrester, and Ting. All three were maritime classics.

Pacing back and forth in the cabin, he told me that he'd been sailing tall ships for years, earning his living as a captain for hire when work was available, doing deliveries and charter trips. Between jobs he was a lobster fisherman in Connecticut coastal waters near his home. He explained that there had been worldwide interest in tall ships in recent years, but America was lagging behind other countries and needed to get more committed if it wanted international respect. He talked about the magnificent "tea clippers" like *Cutty Sark,* that even after the demise of most other commercial sailing vessels in the late nineteenth century continued to race across the oceans to get the best prices on their cargoes of tea. He told me amazing stories of his childhood on the island of Sumatra, where his best friend was an orangutan, and of the many trips his family took by ocean liner to visit relatives in Holland. He had passed through the Suez Canal two dozen times before he was eight years old, twice a year across the Indian Ocean and Red Sea and Mediterranean Sea and North Atlantic, and in all the years since had felt the open sea pulling with an insistence he was powerless to resist. At eighteen he let himself be pulled for good, and would never feel at home on land again. The ocean had magic in it. He described dolphins leaping ahead of the bow wave one moonless night on the Atlantic.

"In the darkness the dolphins appeared black," he said. "But as they swam they were surrounded by phosphorescence that made them shimmer in outline. They looked like leaping constellations. Have you spent much time on the ocean?"

"Very little."

"You can smell land," he said. "You don't realize it while you're living ashore, but after you spend even a few days on the open sea where there's nothing to smell but salt water and the stink of your own body, when you get near land again you can smell the dirt and the trees."

(And that June, after an open-sea crossing of the Gulf of Maine, Hajo and I would stand together on deck as we approached land at midnight, with the full moon straight overhead and the muscular coast of Maine silhouetted against the sky, and catch a sudden warm shore breeze laden with the odors of woodsmoke and balsam—not dirt, because dirt lies thin along that coast, but the products of it—and Hajo, standing alert at the helm, would turn to me and say, "There! Smell it?" as if our conversation had taken place minutes ago, not eight weeks earlier and seventeen hundred miles away.)

Hajo took the job on the *Malabar* because he was curious to see the Great Lakes and eager to know how this boat performed. But before she could be sailed, she needed repairs. The work so far had been grueling. At first he labored alone, but for the past several weeks he had been assisted by a first mate, Matt Otto, who had flown to Traverse City from Baltimore to help with the repairs and delivery. The weather that March and April had been typically cold, and the only heat on the boat came from the diesel cooking stove in the galley. Working twelve-hour days, Hajo and Matt had refitted the rigging, reconditioned the engine, and reconstructed the cabins. But by far the most difficult job was repairing the ferro-cement hull, which had decayed so badly that an inspector hired by the previous owners had condemned the boat and recommended she be scuttled. Hajo and Matt had already patched and rebuilt large sections of hull.

"Where's Matt now?" I asked.

"In the bow. Come and meet him."

I followed Hajo up the ladder, forward across the deck, and down another companionway ladder. Matt was crammed inside a tiny collision bulkhead in the bow of the boat. He extended his arm with difficulty through the access hole and I shook his hand. Later, when I met him again, I would see that he was tanned and fit, a thirty-year-old with a young man's aggressive restlessness. Now I saw only that his hand was callused and covered with cement dust.

Near the bulkhead, wooden berths had been removed from the walls and the hull was demolished. It looked like a bear had gone at it searching for hidden groceries. The cement was ripped away, revealing steel reinforcing rods and a meshwork of chicken wire. Some large sections had already been repaired and were patched with freshly applied mortar.

Though I knew the boat was damaged, I was surprised at its extent. For a dozen years the *Malabar* had offered evening cocktail-and-dinner cruises on Grand Traverse Bay, often with live traditional music performed by a popular local band called "Song of the Lakes." Afternoons during the school year, volunteer instructors used the boat as a floating classroom, taking groups of schoolchildren out on the bay to introduce them to basic limnology and aquatic biology. An entire generation of kids in the Grand Traverse region learned on the *Malabar* to use Secci disks to measure water clarity and microscopes to identify rotifers and copepods and other tiny aquatic animals. Thousands of adults had relaxed aboard her while watching the sun set over the hills of Leelanau County.

But in the last couple years the *Malabar* had been neglected. Water had seeped beneath the teak deck and entered fissures in the hull. That winter I had noticed her resting at anchor in the harbor, her cabin tops piled with snow, her hull listing alarmingly. The local newspaper reported that the owners of the boat had decided the decay in the hull was too extensive to fix and they planned to tow her into the bay and sink her to the bottom, where she would become a divers' attraction. That would have seemed an ignominious end to a boat that had served the community with dignity, so I was pleased when I read a few months later that a businessman from Maine named Steve Pagels had purchased the boat and planned to put her back in service.

Pagels, who already owned several tall ships on the East Coast, had heard about the plight of the *Malabar* and flown to Traverse City to look her over. On the recommendation of an acquaintance he contacted Hajo, who had a good reputation as a surveyor and captain and was experienced in building ferro-cement boats. Hajo flew to Traverse City, inspected the boat, and agreed with Pagels that most of the damage to the hull was superficial, though Hajo was concerned that the steel framework—the ferro in ferro-cement—might be weakened by corrosion. Nonetheless, Pagels made an offer to purchase the boat. The deal went through, and he hired Hajo to oversee repairs and sail her to Maine.

It would complete a circle for the *Malabar*. She was built in 1975 at the Long Beach Shipyard in Bath, Maine, as a replica of the coasting schooners once common along the East Coast. Initially christened *Rachel Ebenezer*, she was put to work doing day charters, first along the East Coast, then in Key West and the Virgin Islands. She was later renamed *Malabar*, and in 1987 the Traverse Tall Ship Company purchased her, sailed her to Lake Michigan, and put her to work in Grand Traverse Bay.

Schooners very much like the *Malabar* had once worked the lakes by the thousands. They and the "packet" steamers of the nineteenth century carried cargo and passengers and were the tractor trailers and Greyhound buses of the inland seas. Larger vessels, like the 350-foot, 288-passenger SS *Keewatin*, which is now moored as a floating museum in Saugatuck, Michigan, were more luxuriously outfitted, with ballrooms, lounges, and elegant cabins that rivaled the cruise ships of the Atlantic and Mediterranean. Around the turn of the twentieth century, at the peak of the excursion boom, thousands of passengers a year embarked from major ports for such destinations as Mackinac Island and the Thousand Islands region of the St. Lawrence River, as well as to dozens of lesser known lakeside resorts and amusement parks around all five lakes. In 1987, the *Malabar* had been at the forefront of a resurgence in passenger cruise vessels in the Great Lakes. In the 1980s and '90s, several cruise lines would launch luxury excursion tours on the lakes. Only one or two were still in operation by the end of 2001.

In an odd coincidence, Hajo lived in Bath, Maine, in 1975 and often visited the boatyard to watch construction of the *Malabar*, née *Rachel Ebenezer*. He would stop by and observe the workers, picking up construction tips, now and then volunteering to help. What he learned he would put to use building his own first boat, his dream since childhood. After reading every book on boatbuilding he could find, he concluded that ferro-cement was the only hull material he could afford. In those days of hippy community spirit, boatbuilders up and down the East Coast converged for "plasterings," the community barn-raisings of the coast, where everyone pitched in to slap cement over steel framework, which is the critical stage of ferro-cement construction. With the experience he gained from several plasterings, including the *Malabar*'s, Hajo built his own ferro-cement boat, a 37-foot sloop he named *Swallowtail*. During the year he spent sailing her up and down the coast of Maine, he got married, and he and his wife, Lee, lived aboard during their first winter together. A coal stove in the galley provided the only buffer against the damp Downeast cold.

So Hajo was experienced with ferro-cement—had experience building this very boat. He showed me what he and Matt had accomplished so far. "We've figured out how to patch her, but it's been very laborious. Most of it's been a bloody lark. Nobody knows much about repairing ferro-cement, and I don't know if what we've done is right. Ask me in ten or fifteen years. My philosophy is to put cement on cement, not epoxy on cement, and it seems to be working."

I said cement seemed a bizarre choice for hull material. It sounded like a stunt, like the cement canoe races staged by engineering students every spring at Michigan State University. Hajo grew adamant.

"Ferro-cement has a bad rap," he said. "People call cement boats ugly boats, but they're being ignorant. If the boats are ugly, it's because the people who built them didn't have the money to finish them properly. But damn it, the hulls are still here. They last. They're durable. Say what you want about steel and wood, but ultimately ferro-cement might prove to be the most redeemable material out there. It's inexpensive. It doesn't rust or rot. It's easy to repair."

Most of the damage to the *Malabar*'s hull was spalling—pieces of the original cement flaking and falling away where water had infiltrated. Hajo and Matt tapped every square inch of cement above the waterline with ballpeen hammers, listening for sounds of decay, then tore out paneling, bunks, and cabinets, and tapped every inch of the interior hull. "Good cement has a lovely resonance," Hajo said. "Bad cement thuds. We sounded the whole boat and marked the thuds with chalk. The worst damage was on the bow. There were places where we could have kicked through it. My first reaction was horror. It was all spalled. It was scary."

Luckily, most of the spalling proved to be in the quarter-inch skim coat. They re-skimmed with high-adhesive cement and experimented with acrylic fortifiers to increase adhesive properties and make the cement more flexible. The macro-epoxy fortifier they finally settled on worked also as a penetrator to help seal the hull.

The work required analytical problem solving of a kind Hajo enjoys. He kept a file of the spec sheets for every product they tested during the trial and error process. He and Matt bent reinforcing rod to match the shape of the hull, attaching mesh wire to it with hog-ring staples and adhesive caulk, then applied bag after bag of their cement/epoxy mix, several tons of it. They rebuilt the hull to be "seriously sturdy," in Matt's words, "like a rock, for our own peace of mind." The consequences of failure were too serious to allow carelessness. "My ass is on the line," Hajo said. "Do you think I want to die on Lake Erie?"

The repairs did not always go smoothly. Once Hajo spent an afternoon standing in the yawlboat, spreading cement above his head onto an over-hanging expanse of the bow. Several onlookers watched from the dock. When he had finished he stepped back, pleased with the job, and the entire structure collapsed, hundreds of pounds of cement slumping and falling into the yawlboat and exploding in the water around it. To make matters worse, he and Matt began to annoy each other. They were living in too close proximity, working too many hours together on a difficult project. Some days the threat of murder filled the air.

But they persisted, learned how to get along, and in time learned how to fix the boat. They discovered that cement when mixed to the consistency of Silly Putty would adhere even to the overhanging bow. They repaired all the decay above the waterline. Below the waterline was another matter. Nobody knew how bad the damage was there, and Steve Pagels was not willing to pay to have the boat hauled out and inspected unless he absolutely had to (it would be necessary in Maine before the U.S. Coast Guard would grant approval for chartering). Below the waterline was the unknown factor. Everything else they put in order. They painted the exterior hull and reassembled most of the walls and berths in the interior. They cleaned and lubricated the diesel engine, a 136-horsepower Ford Lehman. They restored the galley and one of the heads to working order, and were ready to start on the rigging.

Hajo and I climbed the ladder to the deck and looked up at the masts. They stood as bare as telephone poles, and shared their approximate dimensions. All the spars and booms lay piled on the deck. Hajo explained that the boat was 105 feet long (including her bowsprit and the main boom that hung over her stern), displaced 74 gross tons, was gaff-rigged, with topmasts and topsails, could carry 3,000 square feet of canvas, and could sleep twenty-one people. He said he planned to have her ready to sail by the middle of May. I was welcome to join the crew.

Two days before departure, I returned to the *Malabar* and threw myself into preparations. Already aboard was Harold Kransi, a recently retired placement counselor for a technical school in Kalamazoo. Even-tempered, heavyset, always ready to laugh, he described himself as a "gentleman adventurer" and said he was determined to make excellent use of his retirement. Like me, he had volunteered to be a deckhand. He too first heard about the trip from a newspaper article and had called Steve Pagels to offer his services.

Harold and I worked together most of the day rebuilding the forecastle bunks (Hajo and Matt had torn them out of the way to get at the hull) and replacing cabinets and countertops in the forward head. Harold is a former U.S. Navy man and knowledgeable about the lore of the sea. Toilets are called heads, he said, because in the old days sailors relieved themselves from the bowsprit shroud, near the figurehead. We reassembled the cabinets and bunks, then cleaned the forecastle and the small cabin aft of it. They would serve as our primary storage areas during the voyage.

The second day, Harold joined Hajo and Matt on deck putting up the rigging, and I worked alone trying to install the toilet in the forward head. It had been dismantled and stored in the forecastle after Hajo determined that the plumbing didn't work. It didn't work, I discovered, only because it had been winterized with packing grease. I inserted a length of wire with a rag tied to the end and plunged the copper intake pipe. A black wad the size of a cigar shot out the end, followed by a jet of clear water.

Hajo, Matt, Harold, and I spent the night on board, Thursday, May 18, planning an early start Friday morning. Hajo's quarters were in the captain's cabin, attached to the main, aft cabin. The rest of us selected cabins from the six below deck in the midships compartment. One had already been claimed by the fifth and final member of the crew, Tim Smith, who would arrive tomorrow. He had come the day before and decorated his cabin with pictures on the wall, a vase of flowers on a shelf, and heaps of down comforters and pillows on his bunk.

I chose a middle cabin and moved my gear in, tossing my sleeping bag on the bunk and hanging my foul-weather gear on hooks on the walls. I lashed a milk crate to another hook and stacked it with books and personal items. The cabin was small, with a tiny nonfunctional sink on one wall and barely enough floor space to stand in. A small hinged window above the bunk provided a view of the feet of anyone walking past on the deck. My cabin and Hajo's would prove to be the ones that leaked the least in storms.

As night fell, we sat on deck and talked. The evening was clear and turning cold. A few stars showed in the sky, and the moon rose early. Tomorrow night the moon would be full. Everyone was tired and we went to bed before ten o'clock.

At dawn, I woke to the sound of Hajo walking overhead. The morning was bright and cool, the wind light. We did last-minute chores—pumping the bilges, sponging the yawlboat clean and hoisting it to hang securely from the stern of the schooner, lashing to the deck five 55-gallon drums filled with diesel fuel. We stowed loose gear below deck and installed a new Garmin GPS unit in the main cabin and screwed its antenna to the deck railing up top. We chatted with well-wishers who came for a final look at the *Malabar*. Tim arrived about nine, grinning under boxes of groceries. He was tall and lean, a happy Ichabod Crane, fifty-five years old, accustomed to being listened to, bubbling with talk and good cheer. Already he had stocked every storage space in the galley and main cabin with food. He announced we would be eating like kings.

With departure approaching, we faced a problem. For weeks the harbor-master, Gary Hill, had expressed concern about the depth of the water. The Great Lakes were at near-record low levels, a consequence of irregular long-term cycles that might or might not have something to do with global warming, and the harbor was shallower than it had been in decades. The *Malabar* draws eight and a half feet of water, and no matter where Gary sounded bottom he could find it no deeper than seven feet.

Hajo refused to take the problem seriously. It was characteristic of him. He was a 1960s flower child and has kept some of the philosophical rem-nants. Be cool, trust the universe, don't worry about what you can't control. He was confident that everything would work out for the best. His relaxed attitude extended to plans both short and long term. He remained unde-cided, for example, whether we would ultimately leave the Great Lakes through the St. Lawrence River to the Atlantic, or take the Erie Canal to the Hudson River. The St. Lawrence was the longer route and more scenic, but it led north to a latitude where in this season we were likely to encounter icebergs. The Erie Canal is a shortcut, but it would require stepping the masts, not an easy job on a boat the size of the *Malabar*. Hajo's tendency when faced with this sort of dilemma was to postpone it. "Not a big deal," he said. "We can decide when the time comes. In the meantime, the im-portant thing is to get under way." But first we had to get more water under the hull.

A breeze came up from the east, an unusual direction on this coast, where the predominate winds are from westerly quarters. If wind blows from the east it's often the leading edge of a cyclonic storm front, with severe weather to come. But this easterly came fresh on a clear day, with a forecast of more clear days ahead. It blew a two-foot chop across the bay and pushed a surge of water into the harbor.

Such surges are common in the Great Lakes, and are either a "wind setup," caused when a large volume of water is blown to the windward shore, or a seiche (pronounced "saysh"), a word coined in the early nineteenth century to describe the oscillating waves that periodically cross Lake Geneva in Switzerland. Seiches are caused by sudden changes in wind and baro-metric pressure that displace water from one shore to the other, an effect similar to that achieved when you blow across the top of a glass of water, then stop. The water piles up on the far side, but when the wind (or barometric imbalance) ceases, the water is released, bounding back in its quest for equilibrium. The Great Lakes don't contain enough mass to generate measurable tides, but setups and seiches can raise and lower water levels by several feet from day to day or even hour to hour. Seiches in Lake Geneva are often about three feet high. The largest in the Great Lakes usually occur on Lake Erie, where they've been recorded as high as thirteen feet. Usually they're harmless, though sometimes they force ships to wait a few hours before entering or leaving a harbor. But on rare occasions they can be dangerous. When a storm is severe enough, it can create an "edge wave" at the front of the seiche that rushes without warning across the lake. Such a wave killed ten beachgoers in Lake Michigan on the Fourth of July, 1929. A similar incident occurred along the lake's eastern shore on an otherwise calm day in July 1938, when sudden ten-foot waves appeared, washed over the beach, and drowned five swimmers. After a severe thunderstorm in June 1954, a ten-foot seiche struck southern Lake Michigan, washing dozens of people from piers along the Chicago waterfront, killing eight of them.

We stood on the deck of the *Malabar* and watched the water rise on the pilings across the channel. It climbed a foot, a foot and a half. We could have been watching a reservoir fill. Suddenly, just in time for our noon departure, the harbor was deep enough to allow us to leave.

Hajo seemed unsurprised by this good fortune. He started the engine, ran down the companionway ladder to check the gauges, and ran back up to the helm and adjusted the throttle. Someone handed us the dock lines, and Harold, Tim, and I tossed them on the deck. "No, no, like this," Matt said, and made us coil them properly.

Gary, the harbormaster, zipped around the schooner in his Zodiac raft. He nudged the bow at the waterline, bounced off, nudged again, and accelerated his outboard to a whine. It churned the water, and slowly the bow of the *Malabar* swung away from the dock. When the bowsprit pointed into the channel, Hajo shifted the engine into gear and the boat crept forward. We eased over the shallow bottom without touching and entered the channel leading to the outer harbor. Other boats in the marina blasted their air horns in farewell. People on the docks waved and applauded and took photographs. Two television crews recorded the event for that evening's twenty-second nod to posterity. A woman cupped her hands to her mouth and shouted: "Thank you for saving the *Malabar*."

Chapter 3

LAKE MICHIGAN

Heading up Grand Traverse Bay, with a world of freshwater opening before us, it was easy to forget that this was ocean once. Six hundred million years ago a shallow, warm, saltwater sea covered much of Laurentia, the ancient continent that would become North America and Greenland. The sea advanced and withdrew at least ten times during the Paleozoic Era, the age of ancient life, which began with an explosion of marine organisms and ended three hundred million years later with our planet's greatest extinction event. The continent that was to become North America straddled the equator then, and the sea that inundated it was warm and fertile, rich with corals, mollusks, brachiopods, trilobites—a variety and abundance of life forms so profuse that when they died, they settled to the bottom and were compressed by their own weight into beds of limestone hundreds of feet thick.

Fossil remnants of that Paleozoic sea can be found everywhere along the coast of Lake Michigan. The most sought after is the so-called Petoskey stone, a fragment of fossilized coral that when polished by waves or hand reveals a pattern like honeycomb. The deposits of salt left behind by the ancient seas are now mined from beds deep underground. Sand that accumulated on Paleozoic beaches is excavated and sold to industries that prize its cleanliness and uniformity of size.

Before the sea there was only rock. Three billion years ago, during the

Precambrian Era, volcanoes erupted and released magma to the surface. As the magma cooled and hardened, tectonic stresses folded it beneath other rock or lifted it into mountains. In time the mountains eroded and became the Canadian Shield—that broad, low, relatively flat region of exposed bedrock stretching from Canada to Greenland and bordering much of the northern and northwestern shores of Lake Superior. South of there, around the rest of the lakes, the bedrock lies buried beneath sedimentary remnants of the ancient ocean and beneath the glacial debris that came later.

Long after the lava stopped flowing and the saltwater seas receded, a broad circular belt of shale was left at the center of what would become the Great Lakes region. Because shale is soft, it eroded, forming river valleys surrounded by escarpments of hard limestone formed from the skeletons of the creatures that had died in the old ocean. The winters grew colder. Snow fell where once the weather had been tropical. The snow fell faster than it could melt and accumulated, the weight of it compressing the old snow into ice. It accumulated most heavily near Hudson Bay, where the ice grew to be two miles thick and was so heavy that it could not support its own weight and began to flow a few inches a day southward. Eventually the ice moved in sheets hundreds of miles across and a mile or more high.

As the ice advanced, it pulverized slate and other soft rocks and fractured and bulldozed bedrock and pushed it away or carried it along. The ice was diverted from highlands of resistant bedrock and limestone, separating it into lobes that plowed through the river valleys, scouring them deeper. It dredged thirteen hundred feet deep into what is now Lake Superior; nine hundred feet into Lake Michigan. Debris was carried to the edges of the lobes and deposited as the ice melted, forming hills of undifferentiated till— a mix of gravel, clay, sand, and boulders of all sizes.

The first glacier probably came about 1.8 million years ago and lasted a few thousand years before the climate warmed again, the ice retreated, and plants and animals returned. Other glaciers advanced and retreated many times during the Pleistocene Epoch. Geologists for decades counted four major advances: the Nebraskan as the oldest, followed by the Kansan, Illinoisan, and Wisconsinan, named for the states where they made their last stands. Now they think that ten or more advances reached the Great Lakes region, each obliterating most of the evidence of the previous ones. At their peaks, the ice ages had enormous impact on the planet. So much of the earth's water was locked up in ice that the oceans shrank three hundred feet below their current level. The weight of the ice compressed the land two thousand feet lower than it had been. With each retreat of the glaciers, the land would spring back at the rate of a foot or so every century. Some of it continues to rebound to this day.

Scoured deep by the glaciers and sinking beneath the weight of the ice, the old river valleys became reservoirs that filled as the glaciers melted. Those reservoirs became the early Great Lakes. Countless other lakes were formed this way, from the Boundary Waters of northern Minnesota and Ontario, to the Finger Lakes of New York, to Lake Champlain and Walden Pond, to Lake Geneva and the fifteen lakes of the English Lake District. When ice blocked the northern portions of the Great Lakes, water sought to escape south. Lake Superior poured through a valley across the central Upper Peninsula of Michigan and emptied into Lake Michigan; today only the little Whitefish River flows there. With what would become the St. Lawrence River blocked by ice, Lake Ontario escaped south through the Hudson River valley. Lake Michigan flowed southward through what is now the Chicago River and joined the Mississippi. Finally, seven to ten thousand years ago, when the last glacier retreated north, the river we now call the St. Mary's opened, connecting Superior to Lake Huron, and all the lakes found an outlet through the St. Lawrence to the Atlantic.

Like glacial lakes everywhere, the Great Lakes are geological adolescents, especially when compared to the earth's other great freshwater sea, Russia's Baikal, which formed in a rift in the earth's crust twenty-five million years ago. In the grand scheme, the Great Lakes are barely out of diapers.

And they *look* young. Limnologists classify them as oligotrophic, meaning they are clear, cold, and relatively barren of life. Organisms don't thrive in oligotrophic lakes because their bottoms are composed mostly of sand and rock too clean of sediments to allow plants to take root, and because their waters, though rich in dissolved oxygen, don't contain enough nutrients to be highly productive of algae and other organisms. There are exceptions to this rule in the Great Lakes, most notably in bays like shallow and naturally fertile Green Bay, which was named early in the nineteenth century in reference to the algae that periodically tint its waters. A larger exception is Lake Erie. The ecological devastation that struck Erie in the 1960s was caused by an influx of nutrients from fertilizers, detergents, and municipal and industrial waste that essentially made the lake old before its time. Erie's recovery in the last thirty years has been dramatic, though the lake still has serious problems. All five of the Great Lakes have suffered much environmental damage, but their waters remain mostly clear. Portions are as clear today as they were hundreds and perhaps thousands of years ago.

The earliest people to see the Great Lakes probably arrived near the end of the last Ice Age and migrated north as the glaciers receded. Like today's northern tundra of Canada, the land around the lakes was exposed to wind and was grown over with Arctic grasses and thickets of dwarf willow and birch. At about the same time that the first small cities were appearing along the Tigris and Euphrates rivers in Mesopotamia, the people of the Great Lakes were trailing herds of wooly mammoths, muskox, bison, and caribou, which themselves were following the grasses that had begun colonizing the newly open land. As forests succeeded the tundra, wildlife patterns changed and the people changed with them. The Plano culture of the Paleo-Indian period was replaced by the Shield culture of the Archaic period and finally by the Algonquian culture of the Woodland period. By the time European explorers arrived in the early 1600s, native traditions went back five hundred generations.

The explorers who came first were searching for a shortcut to Asia. Later, when it was apparent that no shortcut existed, they came for the wealth that could be extracted from the enormous new continent. The waterway leading to its heart made a convenient highway.

The *Malabar* motored into a moderate northeast breeze. The wind was colder than it had been on land; we went below and put on jackets and hats. Hajo decided not to raise the sails until we became more comfortable with the boat and with one another. The engine throbbed effortlessly and a small wake veed away from our stern.

Matt had announced that one of us must always stand bow watch to keep an eye out for debris in the water and oncoming boat traffic. I stood beside the bowsprit scanning the lake. In the water ahead I saw a small weathered board and watched it slide past the bow and travel slowly the length of the hull and disappear gradually behind us. I wondered how I would find the patience to cross four Great Lakes at this pace.

Grand Traverse Bay is the most prominent bay on the mostly uninterrupted Michigan side of Lake Michigan. It's a deep cleft in the shore, about thirty-five miles from head to foot and fifteen miles wide where it opens into the lake. Early travelers in birchbark canoes avoided the detour around the bay by making an open-water crossing at its mouth. The French fur traders called it *la grande traverse*. A shorter crossing to the north, at the bay in Petoskey, was *la petite traverse*. The voyageurs, like Native American paddlers before them, often camped at the headlands on one side of the bay or the other so they could make the crossing early in the morning, when the wind was likely to be down.

Most of its length the bay is divided into two nearly equal halves by Old Mission Peninsula, a twenty-mile-long and two-mile-wide finger of rolling hills covered with cherry orchards, vineyards, woodlots, and houses. The tip of the peninsula touches the 45th parallel, the halfway mark between the equator and the North Pole. At the foot of the peninsula is Traverse City, established about 1840 as a lumber town, but long since grown up into a resort metropolis of twenty thousand or so, with another forty thousand living in the surrounding county. The economics of the region are complicated. Tourism is the main industry, which means that seasonal jobs abound but not many pay well. People who abandon careers in Detroit or Grand Rapids to move to northern Michigan must often make significant financial sacrifices. We have an expression: "A view of the bay is worth half the pay."

Until recently, our local economy depended upon agriculture as much as it did tourism. The bays are near the northern limit of Michigan's fruit belt, which extends for a couple hundred miles along the shore. The wind off the lake moderates the climate, making summers cooler and winters milder than they are even a few miles inland. South of the Grand Traverse region, apples, peaches, and grapes predominate. Around the bays, especially on the Leelanau and Old Mission peninsulas, cherries are the most common crop. Traverse City bills itself as the "Cherry Capital of the World," though more of the fruit is now grown elsewhere and a fickle market increasingly prompts area orchardists to uproot their trees and replace them with grapevines.

The lake here, as along much of the Great Lakes, is bordered by rolling hills that were pushed up long ago by the glaciers. Even now, ten thousand years later, you can see where meltwater once surged between the hills, carving valleys, and where debris was dumped from the ice to form the low hills called moraines. Look at the land long enough and you begin to realize that it was devastated by glaciers. They ripped it apart and left it a wasteland. Bulldozing forward a few inches or a few feet a day, they eventually departed and left heaps of rock, gravel, sand, and clay behind. Over the centuries, water and wind smoothed the edges, and vegetation covered most of the land with raiment. But you can still see the bones and sinews beneath.

You can also see the terraced levels of former beaches, where the water stood fifteen or thirty or fifty feet higher during various times in the past few thousand years. The terraces are easiest to see from the lake, especially when leaves are down in November and a dusting of snow highlights the ground's contours. For the last four thousand years most of the region has been covered with forests of pine, spruce, hemlock, oak, maple, aspen—at least until loggers arrived at the end of the nineteenth century and decimated them. Now the shore is cloaked with second- and third-growth woods.

From below deck, Tim shouted that lunch was ready. We would eat in shifts. Matt, Tim, and I went first; then Hajo and Harold. Tim, grinning, met us at the bottom of the stairs. He wore an apron and had draped a towel over his arm—the chef of *Chez Malabar*. On the table sat plates of Greek-style egg salad, loaves of French bread, slices of tomato seasoned with basil leaves and feta, and asparagus spears. On every plate was a sprig of rosemary, and in the middle of the table sat a vase of flowers. We took our seats and dug in. The meal generated a higher level of optimism.

Tim lives in Traverse City, where he once owned a successful interior furnishings business. A few years ago he wrote a children's book entitled *Buck Wilder's Small-Fry Fishing Guide*. It proved wildly popular, and he now works full time writing and marketing a series of Buck Wilder books. Like Buck himself, Tim is irrepressibly cheerful and energetic. As we ate, he told me that he learned about the *Malabar*'s upcoming journey to Maine while getting his hair cut at Robertson's Barbershop in Traverse City. A scruffy guy sitting in the chair next to his had a wild tale to tell about a pirate ship he was sailing to the ocean. The scruffy guy was Hajo. By the end of the haircut Tim had signed on as cook.

After lunch, we gathered on deck. Matt took bow watch; Harold and Tim and I sat on the deck rails near Hajo at the helm. The *Malabar* rode easily over a two-foot chop. The wind remained light and chilly from the northeast, under a bluebird sky that promised days of good weather ahead. May is a fine month to be on the Great Lakes. Nights are cold, but the days are mild and storms are less common than in the summer and autumn. We looked out across the lake, talked, looked some more.

Hajo said quietly that perhaps it was a good time to discuss a few rules of the boat. First and foremost, the captain's word is law. No arguments, no appeals. Second, if you don't know how to do something, ask. There's no shame in not knowing, but harm can be done if you try to bluff your way through. Third, no consumption of alcohol while under way. Save it for the dock. Fourth, no pissing from the boat. Most male drowning victims are found with their zippers open; use the head. Fifth. Well, maybe there wasn't a fifth. The fifth rule was to remember the first four.

At the mouth of Grand Traverse Bay the horizon spread out before us and we entered Lake Michigan. The name of the lake is derived from the Algonquian word *michigami* or *misschiganin*, meaning "big lake." Europeans knew it first as the Lake of the Illinois, named for the people who lived along its southern shores and who were eventually overrun and nearly annihilated by the Iroquois. It's the only Great Lake that does not share a border with Canada.

We followed the coast north, past the resort town of Charlevoix, named for the French Jesuit who coasted these shores in a canoe in the 1720s. Even staying close to shore and with the wind light, it was obvious that we were on a sizable body of water. With a length of 307 miles, a maximum width of 118 miles, and a surface area of 22,300 square miles, it is the third largest of the Great Lakes, and the sixth largest freshwater lake in the world. And it's deep: the average depth is 279 feet, and the deepest spot is 923 feet.

We passed Little Traverse Bay. Tucked away at the head of it are the small cities of Harbor Springs and Petoskey, at this distance appearing as not much more than a bit of glitter and some geometry on the hills. With binoculars we saw rows of tiny condominiums at Bay Harbor, said to be the largest resort complex in the Midwest.

Students of twentieth-century literature know this shore as Hemingway Country. Though he was born (in 1899) in the Chicago suburb of Oak Park and went to school there, the young Ernest Hemingway spent his summers at the family cottage on Walloon Lake, a few miles from Petoskey. Those summers on the lake provided much of the raw material he would shape into his early stories. In Paris in the 1920s, he pinned a map of northern Michigan on the wall above his writing table and set to work with the intention, as he later described it, of making every word count, making every detail add to the overall structure of a story. He wrote about the places he remembered: Walloon Lake, Lake Charlevoix, Horton Creek, and the rivers—the Sturgeon, Pigeon, Black, and, especially, the Fox, which would be transformed into the Big Two-Hearted. The words he laid down on the page brought to life the waves running up the beach, the burned-over hills, the stands of pine with the wind high in their branches, capturing northern Michigan more truly than anyone has. In a fragment unpublished until it appeared in a 1972 collection, *The Nick Adams Stories*, Hemingway wrote: "He, Nick, wanted to write about country so it would be there like Cézanne had done it in painting. You had to do it from inside yourself. There wasn't any trick. Nobody had ever written about country like that. He felt almost holy about it. It was deadly serious. You could do it if you would fight it out. If you'd lived right with your eyes."

Years ago I knew an old man named Bud Schulz who had lived in Petoskey all his life and was a child when Hemingway was there. He remembered the talk around town after the young author began making a name for himself. A few of the stories in the first collection, *In Our Time*, were scandalously suggestive of sexual matters, inciting some locals to refer to Hemingway as "Dirty Ernie," though as time passed and his fame grew they became more tolerant. Bud was friends with Hemingway's younger sister, Sunny, who continued to live in Windermere, the old Hemingway cottage on Walloon Lake, until her death in 1995. Bud, who at age eighty thought nothing of throwing an aluminum canoe on his shoulders and humping it down the bank to the Bear River, urged Sunny to invite me to the lake to go canoeing with her. She cordially declined. It was well known by then that she had been harassed so often by Hemingway worshipers that she refused to talk about her brother at all, and she probably assumed I was mining for family secrets. It was a bad idea, anyway. She was in her eighties by then, and Walloon can get choppy.

After Sunny's death, her son Ernie Mainland inherited Windermere, where he lives today with his wife, Judy. Not long ago Ernie invited several authors for a private tour of the cottage. He's a stout, red-faced, robust man in late middle age who bears a striking resemblance to his uncle. He met us on the dock—we arrived on a pontoon boat from across the lake—and shouted, "This is not the Hemingway cottage! The Hemingway cottage is down the lake!" He laughed at our perplexity and bellowed, "This is the right place! Get your asses ashore!"

He served us wine and hors d'oeuvres and showed us around his home, pointing out his uncle's ancient baitcasting rod and reel hanging on a wall and the hundred-year-old landscapes painted by Hemingway's mother, Grace Hall Hemingway, who signed them "Hall Hemingway." We examined a door frame where incremental pencil lines and dates recorded the growth of young Ernie and his siblings, and studied a map of Walloon for landmarks mentioned in the Nick Adams stories. Ernie Mainland explained that he had not been aware of his uncle Ernie's celebrity or his impact on American letters until he was in college and learned he could trade on it to his advantage. "It got me a few Bs when I deserved Cs," he said. He admitted without apology that he had read only one book of Hemingway's short stories (and couldn't remember which one) and had never cracked the covers of *The Sun Also Rises, A Farewell to Arms,* or any of the other novels.

Ernie invited us to take turns sitting for photos in the old two-seater out-house, where ERNEST HEMINGWAY SAT HERE had been painted on the door.

To the northwest, land emerged on the horizon, low and darkly wooded, growing larger as we approached. This was Beaver Island, Lake Michigan's largest island at fourteen miles long and seven wide, home to about 450 year-round residents and four times that many in the summer. Long before the island was a summer getaway it was the scene of a bizarre episode that culminated in its being declared the only independent kingdom in U.S. history. The ruler of the kingdom was a renegade Mormon named James Jesse Strang, who'd been born in 1813 in the northern New York village of Scipio. Strang became a lawyer and a teacher and migrated to Wisconsin in 1843, where he converted to Mormonism. A year later he visited church founder Joseph Smith at the Mormon capital of Nauvoo in western Illinois, and was named a Saint of the church. When a mob of Gentiles murdered Smith later that year, Strang stepped forward and claimed that God had spoken to him in a vision and declared him the new ruler of the Mormons. Church elders disputed his claim and excommunicated him, and the church became factionalized. Most members joined Brigham Young and followed him to Utah to found Salt Lake City, while others took the side of Joseph Smith, Jr., and broke away to form the Reorganized Church of Jesus Christ of Latter Day Saints, a denomination still active today. A few hundred others were convinced that James Jesse Strang had experienced an authentic vision and in 1847 followed him to remote and sparsely populated Beaver Island. They drove off a handful of previous settlers, mostly Irish immigrants who supported themselves by fishing and selling firewood to passing steamships, and began construction of a utopian community.

For a few years the colony thrived. By 1849, fifty families had settled around the harbor at the island's north end, forming a town they named St. James. Strang ruled absolutely, forbidding polygamy among his followers but taking five wives for himself, and handpicking a cabinet of trusted elders. He traveled to Washington, D.C., and lobbied successfully to have ownership of Beaver Island granted to him and his followers. Along the way he gathered converts. During his travels he was accompanied by a prospective wife, eighteen-year-old Elvira Field, who to avoid scandal dressed as a boy, posed as his nephew and secretary, and went by the name of Charles.

Back on the island, Strang claimed to have another vision. This time God disclosed the whereabouts of an ancient holy text, *The Book of the Law of the Land* that named Strang "King of all the Earth," and awarded all the islands of the Great Lakes to his church. His coronation took place on the shore of Beaver Island in July 1850, with four hundred followers as witnesses. The new king quickly established settlements on nearby islands and the mainland, diversifying his influence. He established northern Michigan's first newspaper, *The Northern Islander,* and put it to use promoting his mission. In spite of federal charges that he had overstepped his authority, he campaigned successfully in his legislative district, which included the mainland town of Charlevoix, and was elected to the Michigan house of representatives. He maneuvered to get St. James named the seat of government in Emmet County, giving him power over the hostile Gentiles eighteen miles away on the mainland. He was fearless and tireless, but everywhere he went and with every accomplishment, he made enemies.

They caught up to him in 1856. Two of his own flock ambushed him on the dock in St. James and shot him in the back of the head. Mainlanders immediately sailed to the island and drove the Mormons away, stealing their possessions and claiming their property. The Irish settlers returned and for the next hundred years the island was inhabited mostly by commercial fishermen.

Beaver Island slipped away to the stern, and the sun dropped below the horizon. The evening quickly turned cold. I had volunteered to take the midnight to 4:00 A.M. watch, so I went to bed at eight o'clock and read for a while in my sleeping bag. The boat rocked gently and the engine made rhythmic sounds through the hull.

I woke in darkness and turned on a flashlight to check my watch. Twelve-thirty. Late by half an hour. Nobody had come to wake me, and I was afraid that Matt's thoughtfully planned watch system was already falling apart. The plan was for Matt and Hajo to operate on six-hour shifts, one of them always at the helm. Tim was exempt from night watches because he stayed busy all day cooking and cleaning in the galley, so Harold and I would alternate four-hour shifts through the night. But those first few nights whoever was fresh took over the watch while the other slept.

I climbed to the deck, into cold night air, and everywhere I looked was water flowing silver with moonlight. The moon was full. It hung high between the masts, throwing crisp shadows across the deck. Shore showed as a low black band to the east.

Harold, grateful for relief, was ready to go below and sleep, but first he walked me through the duties of night watch. He and Hajo had worked out a routine that we would follow for the remainder of the trip. Once each hour whoever was not at the helm went forward with a flashlight. He climbed down the steep narrow ladder to the forecastle, raised the hatch at the bottom of the ladder, and shone his light inside to see if the level of filthy water in the bilge was higher than it had been an hour ago. He repeated the procedure at midships and again in the main cabin at the stern of the boat. The gauges on the wall beside the galley must then be checked, and the data entered into the logbook: the exact time of inspection, the location (longitude and latitude in minutes and seconds from the readout on the Garmin), the speed in knots, and our heading. The voltmeter was crucial. If it indicated weak batteries—and it often did, since the running lights drained them every couple hours—the gasoline generator on the forward deck had to be started and run until the batteries were charged.

I finished my first entry in the log, and Harold wished me goodnight and stumbled away to his berth. I returned to the deck. Hajo held the helm while sitting on the small wheelhouse. He was bundled in clothing, the hood of his jacket cinched tight around his face. "There's coffee below," he said.

"Want a cup?"

"Sure."

I climbed down to the galley and filled two cups from a pot Tim had left simmering on the stove. I discovered I could carry just one at a time up the ladder.

"Nice night."

"Glorious."

Glorious, but cold. Winter still lingered in these waters, two months after the ice left. Most of the lakes stay open all year, but this northern portion of Lake Michigan usually freezes over in January and February. I wore several layers of capilene and fleece, a sweater and an overcoat, a stocking hat and gloves, and still wished I had dressed more warmly. I whistled a one-note, "boy-it's-cold" whistle. A mistake.

"Never whistle on a boat!" Hajo snapped.

I had already learned that he was superstitious. It worried him that we began our trip on a Friday, on a boat whose name was once changed. Both are considered unlucky in seaman's lore. While we were still in Traverse City a friend had given him a Sacagawea dollar for luck; this morning, as we were about to leave the harbor, Hajo slapped his pocket, said, "Where's my coin?" and bounded down the ladder to his cabin. When he returned, he said, "I go *nowhere* without Sacagawea," a promise he would honor the entire journey. On deck he always wore his *Annie* hat, a souvenir from a New Haven oyster boat that he admired and considered lucky. Above the door of his cabin he had nailed three wishbones from chickens. There too was a needlepointed representation of a gnomelike creature, the Dutch equivalent of a leprechaun that according to legend came to the New World with the first Dutch explorers. When I asked Hajo about it, he said, "They bring good luck if you see them. I'm positive I think I almost saw one when I was a kid, and I know for a scientific fact that they're in Michigan." His eyes twinkled. He asked if I'd noticed the horseshoe nailed to the samson post in the bow of the boat.

"I noticed it back in the harbor."

"It'll help us make the trip safely," he said with certainty.

He was less certain about another horseshoe, one with a reputation for bad luck. For two centuries Great Lakes mariners have feared the "horseshoe" of the lakes—the route around Michigan's Lower Peninsula, from the bottom of Lake Michigan to the bottom of Lake Huron. It is considered bad luck because the horseshoe shape is inverted, allowing the luck to pour out. Nobody has compiled a complete list of all the shipwrecks in the Great Lakes, but hundreds, certainly, and probably thousands have gone down along that upside-down horseshoe.

Ahead of us waited one of the most treacherous places in the lakes: Gray's Reef, a channel through a complex of submerged shallow reefs composed of the mineral dolomite, which was formed from precipitates of marine animals that lived in the ancient seas. The reef is an extension of the Niagara Escarpment, a geological anomaly extending from Wisconsin's Door Peninsula, around the top of Michigan, across to Ontario's Bruce Peninsula on Lake Huron. From Lake Huron, it cuts east across the arrowhead of Ontario to the land bridge that separates Lake Erie from Lake Ontario, where the Niagara River falls off its edge.

Hajo had mentioned several times already that he was worried about Gray's Reef. On the charts it's a cluster of buoys and lighthouses, with safe passage only through a narrow channel. Hajo knew we would be passing through at night and that we might confront a bottleneck of lake traffic there.

We noticed lights blinking on the water to the north.

"Is that Gray's Reef?" Hajo asked.

"I think so."

"Have you ever been through here?" he asked.

I admitted that I had passed through one night the previous summer during a sailboat race. I tried to add a caveat: "But it was a dark night and I'd hardly slept for three days and I'm not sure I remember . . ."

"But you were here."

"Yes."

"Good. You navigate."

I knew nothing about nautical navigation, but Hajo was so in need of assistance that I didn't have the heart to disappoint him. I went below to the main cabin and with a flashlight examined the charts spread on the table. This end of the lake is spotted with islands and reefs and the long arm of Waugoshance Point. I've walked on the Point; it's a low, rocky nesting ground for gulls, lined with reeds that sway with the waves, and is uninhabited by humans. Gray's Reef is an extension of the Point—a string of small, rocky islands with shallows between. The channel is twenty-five feet deep at low-water datum, and marked at its south and north ends by automated lighthouses perched on carapaces of rock. Between them are double strings of buoys and bell buoys leading the way through the safe water.

Much of that I remembered from last summer's Chicago-to-Mackinac Race, when the sailboat I was on entered the channel with a pack of others. I remembered also that as we approached the reef from the south, we were confronted with a confusion of flashing red, green, and white lights, and no obvious way through them. There was not much else to see. The reef was submerged; the little islands and Waugoshance Point were out of sight in the darkness.

I studied the chart until I could decipher the abbreviations beside each buoy and lighthouse. The first to watch for was the white beacon on the lighthouse at Ile Aux Galets, better known as "Skillagalee," which flashes every six seconds. We had to keep it on the starboard side, maintain a northeast heading, and aim for the first pair of buoys in the series, a red one on the port side, flashing at ten-second intervals, and a green one on the starboard, flashing every four seconds. Once we identified those lights, we could continue our northeast course until we lined up with the second set of red and green buoys. At that point, we would adjust our course to due north. Then the channel should open before us, lined with reds on the left and greens on the right.

Hajo called down for me. "I need to know what to do," he said.

I returned to the deck. With the moon so bright, the entire lake seemed visible—immensities of silver water in every direction—with nothing else to see but blinking clusters of navigation lights ahead. I told Hajo what to look for. We passed the first lighthouse and identified the buoys ahead. Hajo's eyes jumped from the buoys to the compass to the buoys. The compass was housed in a domed binnacle mounted on the top of the cabin forward of the helm. Its internal light was broken, so Matt had earlier clipped a penlight to the brass housing. It gave off a red light bright enough to illuminate the marks on the compass but did not affect our night vision.

We approached the first buoy and Hajo steered the boat north. To my relief, the channel opened before us. On the chart the passage is marked as three thousand feet wide, a little over a half mile, but in the darkness it seemed narrower. If we had been sharing it with a barge or freighter, it would have seemed much narrower yet.

We passed through, counting buoys as we went, a foghorn bleating somewhere ahead, though there was no fog that night. The beacon of White Shoal Lighthouse flashed high on our port side, and we knew we were finally at the top of Lake Michigan. We steered hard to starboard until our bowsprit swung straight east, toward the Mackinac Bridge. The bridge sat low and bright in the distance, a graceful spray of lights across the Straits of Mackinac.

It was exhilarating. I felt like a useful member of the crew.

A year earlier, during the race from Chicago, I came around this same corner on a sailboat and saw the bridge for the first time and felt similar exhilaration. That night had been very dark, no moon or stars, and the wind was strong—much different from tonight. The two boats differed also. The *Malabar* was a large, heavy schooner that plowed through the lake, its diesel engine chugging resolutely. *Gauntlet* was a sleek 44-foot racing yacht, efficient and fast. In the strong wind she had hurtled across the tops of waves, almost silent beneath her load of sails.

Hajo wanted to talk about how the *Malabar* was performing so far. He liked the way she responded to his wishes. He thought she seemed dependable and strong and equal to the challenge of the trip. Again he wondered why the previous owners had sold her. But he was being a romantic, not a businessman. "This boat has a beautiful soul," he said emphatically, daring me to contradict him. When I said nothing, he said, "I think she's grateful to be alive."

Chapter 4

LAKE MICHIGAN

The Chicago Lakefront ◆ Racing to Mackinac ◆
Dunes and Sand Mining ◆ Across the Finish Line

One Friday in July, the summer before I joined the crew of the *Malabar*, I drove to Chicago to meet a guy named Bob on his sailboat. I'd never met Bob, but he had invited me through a mutual acquaintance to ride along with him in the Chicago-to-Mackinac Sailboat Race, which at 333 miles, the full length of Lake Michigan, is the longest (and longest held) freshwater regatta in the world. Bob promised that if I showed up in Chicago the evening before the race, he'd buy dinner and book an extra hotel room for me and in the morning we'd go out on the lake and crush the competition. It was a good plan, I thought. But it didn't work out the way I expected.

From where I live, Chicago is a six-hour drive down the Michigan coast of Lake Michigan. Most of the way the road is two lanes, through woodlands and orchards, past lakeside towns that started as lumber camps and fishing villages and evolved into weekend getaways with time-share condominiums on the beach and motels named "Beachcombers" and "The Breakers" and restaurants designed around maritime themes. As the road approaches Chicago it becomes a highway, of course, and then a superhighway. It's six or eight lanes wide by the time it passes through the city's borderland of suburbs, malls, and franchise businesses; through the warehouse districts and train yards and smoking industrial complexes; past the housing projects, gated communities, and brownstone town houses. At the core of the city rise the skyscrapers of the Loop. But Chicago's heart is in its waterfront.

The Windy City's love for Lake Michigan is nearly as old as the city itself. What began before 1700 as a fur-trading post at the mouth of the Chicago River, was by 1830 platted as a townsite of forty-eight blocks. Soon, settlers began pouring into the Great Lakes and to the plains beyond, making Chicago a hub of transportation, radiating waterways, roads, and railroads. As the town grew into a city, demand for its water frontage increased.

By then, the shore at virtually every Great Lakes city was cluttered with factories, mills, refineries, warehouses, railroads, and docks—the water's edge fouled with ash, slag, garbage, and human waste. Most cities would rise facing inland, their backs huddled against the lakes. But Chicago was different. It would face the lake proudly and maintain a long crescent of beach at its center free of the usual industrial scourges. The credit for that must go to three farsighted men, William Thornton, William Archer, and Gordon Hubbard, who in 1836 had been authorized by the state of Illinois to supervise the sale of canal lands in Chicago. This was an important job and a powerful one, with enormous opportunities for personal profit. But these men were apparently not profiteers. They recognized the economic importance of linking Lake Michigan and the Mississippi River through the Illinois & Michigan Canal (which would be completed in 1848, and replaced half a century later by the Chicago Sanitary & Ship Canal), but they foresaw also the necessity of managing the city's land for the future. They decided among themselves that the Chicago lakefront should be preserved for public use, not parceled off and sold, as speculators were demanding. In case anyone doubted their intentions, they oversaw the printing of a real estate map of the city upon which the lakefront was labeled: "Public Ground—A Common to Remain Forever Open, Clear, and Free of Any Buildings, or Other Obstructions Whatever." A good deal of the shore has remained open and clear to this day.

After Chicago's Great Fire in 1871, city planners began rebuilding with stone and steel instead of lumber, and though they honored the charter to keep the lakefront open, they weren't quite sure what to do with it. It was left, two decades later, for the architect and builder Daniel H. Burnham to make the waterfront the focus of a revolutionary city plan. Burnham soon gained national prominence as the builder of the magnificent "White City"—an idealized, neoclassical model of urban design constructed in Jackson Park for the World's Columbian Exposition of 1893. He went on to put many of his ideals to practice as one of the lead architects in Chicago's civic renewal. At the time of the city's most rapid growth in the early 1890s, Burnham led the effort "to bring order out of the chaos incident to rapid growth," as he explained it. In the process Chicago's waterfront was transformed into one of the finest in any city in the world.

Today along Lake Shore Drive, from Jackson Park north to Lincoln Park, are fifteen miles of nearly continuous parks and public beaches, ball fields, tennis courts, golf courses, and marinas, all linked by jogging and biking trails. Some of Chicago's finest attractions—Adler Planetarium, the Field Museum of Natural History, Shedd Aquarium, Navy Pier—are more interesting for being at the edge of the lake.

That Friday I walked the shore of Grant Park, watching the waves break against docks and jetties. The wind off Lake Michigan smelled fresh and complex, like springtime. Around Chicago the water is more turbid than up north, and you'll often see beards of bright green algae clinging to stones along the shore—evidence of excessive nutrients—but as city water goes, it's quite clean. It wasn't always that way. Until 1900, when the flow of the Chicago River was reversed, raw sewage and industrial contaminants were dumped into the lake not far from the intake pipes through which drinking water was drawn. The river was the city's sewer, and when it contaminated the drinking water it spread disease. One nineteenth-century cholera epidemic reportedly killed one in every twenty Chicagoans. Reversing the flow solved that problem, but of course created others. The Chicago River and its sewage now flowed south to the Illinois River and on to the Mississippi. It would be decades before sewage treatment would clean the river, though today's downstream residents are justified in arguing that it is not yet clean enough.

The Monroe Street headquarters of the Chicago Yacht Club sits at the approximate center of the Chicago waterfront. Some people claim it's at the center of the freshwater sailing universe as well. The clubhouse is a low, unassuming building, practically in the shadow of downtown skyscrapers, bounded by Grant Park on one side and Monroe Harbor on the other. The day I visited, it was a busy place. White-coated restaurant staff whisked past in golf carts, delivery trucks backed up beeping to the service entrance, luxury cars and limousines pulled to the curb at the main entrance and dropped off casually dressed frolickers. Around the front of the building, docks branched away into the harbor, each crowded with sailboats of every size and style, many of them rafted three and four deep. Their masts formed an aluminum forest covering many acres.

I searched the docks until I found Bob's boat, a fine-looking 33-foot sloop tucked between larger boats. A few men in their thirties and forties sat on deck drinking cocktails and laughing. I introduced myself and asked for Bob. One of the men said, "That's me." He was fit, tanned, about my age. I started to ask permission to come aboard and stow my bag, but he held his hand up to stop me.

"There's been a change of plans," he said. "An old buddy of mine showed up this morning. I gave him your spot."

I reminded him that I'd driven to Chicago at his invitation.

He shrugged. "Ain't that a bitch," he said.

Bob was that kind of guy.

But hell, I thought, so what? This was his world, not mine. He owned the boat, he could decide who sailed with him. I'd be just as happy to spend a couple days in Chicago, then head home and go fishing.

But our mutual acquaintance, Dave Gerber, felt responsible for what happened and made it his mission to get me on another boat. Dave is a thirty-year-old sailor and professional sailmaker who lives not far from my home in northern Michigan and has sailed many times in the Chicago Race and others around the world. He spent all Friday evening looking up people he knew at the docks and asking them if they had room for one more on board. Nobody did.

He continued the search Saturday morning, but by then I had little hope he would be successful. Friday night I'd noticed a fit-looking young man in cargo shorts, T-shirt, Teva sandals, and Oakley wrap-around sunglasses—the virtual uniform of sailors young and old—walking from boat to boat and inquiring in a low chant, "Need crew? Need crew?" This morning I watched the same ritual performed by a woman perhaps twenty years old. She was bouncy, blond, athletic, bursting with assurance, and flat-out beautiful. At every boat she received a long head-to-toe appraisal followed by a firm "Sorry." I knew I didn't stand a chance.

So I walked the docks, admiring the boats. They were splendid. Some were wooden, but most were combinations of wood, steel, fiberglass, aluminum, and carbon fiber. There were racing boats and cruisers; sloops, cutters, yawls, and ketches; monohulls, catamarans, and trimarans. The larger vessels, sixty- and seventy-footers, were impressive and sometimes breathtaking, but often in a store-bought way. I was more taken by the smaller boats, the thirty- and forty-footers. To my eye they possessed more individuality than the large boats and had been cared for more lovingly. Many had acquired a well-buffed ambience of the sort you sometimes see in classic cars and motorcycles. On their sterns were painted their names and calling ports. They were from Chicago, Milwaukee, Detroit, Cleveland, Toronto—and from Annapolis, Montauk, Norfolk, Pensacola.

Docked front and center before the clubhouse was Steve Fossett's sixty-foot America's Cup contender, *Stars & Stripes,* a high-tech, cutting-edge catamaran that might have been designed by NASA for skimming across the nitrogen seas of Ganymede. Fossett, who lives in Chicago, was fresh off his most recent attempt to circle the earth in a balloon, and was the heavy favorite to reach Mackinac first. In 1998 he set the Open Division overall record with an elapsed time of just under nineteen hours, which was seven hours quicker than the previous record and twice as fast as most of the better finishes in years when the wind is good. Poor wind can result in excruciatingly slow times. The fastest boat in 1905, for example, took ninety-four hours to reach the finish line.

I stepped onto the clubhouse porch. People sat around tables there, wearing jackets against the cold, drinking coffee and talking about the wind. It was blowing at twenty to twenty-five knots. I looked beyond the harbor and saw waves detonating against the breakwall. "Bumpy out there," someone said.

Now I couldn't remember my seasickness lore. Was it better to have a full stomach or an empty one? But I was hungry, and a sign on the door of the clubhouse dining room offered a buffet breakfast for a good price. I walked in, collected a plate, and went at it. Scrambled eggs, sausage, pancakes, fruit, coffee. A second helping of eggs and some hash browns couldn't hurt. I needed more coffee because I had hardly slept the night before. The caffeine made me jumpy, so I ate a few muffins to settle my nerves.

Dave poked his head in the door.

"No luck yet," he said. "But don't give up."

"You've been searching all this time?"

"Yes."

"Have you had breakfast?"

"No. It's better to have an empty stomach."

The Chicago Yacht Club's first race to Mackinac Island was held in 1898, and, with the exception of a few years during the two world wars, it has been repeated every summer since. It remains the premier race in the Great Lakes, although many sailors will remind you that there are two "Macs." The other is the Port Huron–to–Mackinac, which runs the length of Lake Huron, about 250 miles, and is nearly as venerable as its sister regatta on Lake Michigan. Each event draws 250 to 300 boats and about 3,000 sailors, and usually requires two or three days to complete. They're staged on consecutive weekends in July, Chicago taking precedence one year, Port Huron the next.

I entered the bar, looking for Dave. Though it was not yet noon, the place was packed, with men mostly, though a few stunningly dressed women seemed to be trolling. People of color were conspicuously absent. Deeply tanned men stood six deep at the bar and gathered in clots of bonhomie around the room. Most wore khakis or the standard shorts, Ts, and Tevas, and spoke in bold variants of Urban Midwestern, though a few wore blue blazers and talked like Thurston Howell the Third. One guy trumpeting his business triumphs wore a T-shirt that read: OUR DRINKING TEAM HAS A SAILING PROBLEM. Everyone spoke at once and loudly, as if they owned the place. Rarely have I seen so many alpha males under one roof.

Yacht racing is an expensive sport—perhaps the most expensive. It requires big money to purchase the boats, and almost as much to outfit and maintain them. Owners can spend tens of thousands of dollars in preparation for the Mackinac races. The entry fee is insignificant—up to a few hundred dollars, depending on length of boat. A great deal more is spent on each of the twelve or fifteen new sails needed for the race, a different one for every wind condition, plus spares. Lines are replaced, winches overhauled, electronic navigation systems checked by technicians. The morning of the race, divers are hired at the dock to inspect hulls and rudders for damage (or sabotage: pranksters have sometimes cemented toilet plungers to rivals' hulls, causing drag that slows them a knot or two). Crews must be transported to the race and fed and housed, galleys must be stocked, safety equipment must be inspected and upgraded. It's no wonder the sport attracts aggressive, financially successful people.

The clock was against us now. Dave was bowman on one of the elite contenders, *Flash Gordon IV*, a Farr 40 owned by the Chicago architect Helmut Jahn, and needed to be on board by twelve-thirty sharp. Jahn is very serious about racing and does not tolerate tardiness. I began thinking about going home.

Then Dave was running toward me down the dock. It was twelve-twenty-five. "Let's go, let's go. I found one."

I grabbed my bag and ran after him. "One of their guys just canceled," he said over his shoulder. "His wife broke her leg or something. The boat's a beauty, you'll love it. But hurry, hurry."

He stopped behind a sleek racing sloop with several people on deck. Inscribed on the transom was the name *Gauntlet.*

"One thing," Dave said. "I sort of told them you were experienced. I mean, I embellished a little. You can tell them the truth if you want to. But you don't have to tell them *all* the truth. And you don't have to tell them until you're on the water."

Three or four times in my life I had set foot in sailboats. As a child, growing up on an inland lake, I spent my days canoeing, fishing, water-skiing. Sailing was boring, I thought. It was a diversion for the bored rich kids at the other end of the lake.

We went aboard. "This is the writer I told you about," Dave said, to no one in particular. He turned to me and said, "The owner's name is Guy. Introduce yourself when he shows up."

And he was gone. The people turned away and went back to handing bags of gear below deck. A middle-aged man I would later learn was nick-named "B.B." stepped up to me, got in my face, and said, "Are you good? We don't need any deadweight on this boat."

"I need to find Guy," I said.

"He took the truck to get gasoline."

"Where can I wait for him?"

"Parking lot. Look for a white van."

I sat at a picnic table. A van pulled in and the driver got out, walked to the back, and lifted out a five-gallon fuel can. Guy Hiestand was about fifty years old, sun-browned, dressed in shorts and a sloppy T-shirt. I introduced myself.

"Yeah, Dave told me about you."

"What'd he tell you?"

"Said you were a writer looking for a boat. Experienced sailor. Capable guy."

"Some of that's true."

"Which parts?"

"Writer, fairly capable."

"But you're a sailor, right?"

"No."

"You're not a sailor."

"No."

Silence.

I gave him my best pitch. Been around water all my life. Hard worker. Fast learner. Tell me something once and I won't forget it. If I screw up, yell at me, I don't care, I want to learn.

"Okay, okay, you're in. Stow your bag below and give B.B. a hand hauling groceries."

Ten minutes later we threw off the dock lines and motored slowly out of the marina. Even inside the breakwall the water was choppy. Outside its protection the waves came steep and quick, four- to six-footers topped with whitecaps. And building. Within a few hours they would be sixes to eights.

Guy ordered the mainsail and jib raised. Our crew was nine experienced sailors—and me. Most of the others had sailed together many times and all of them knew what they were doing. I knew just enough to stay out of the way.

We began circling near the big Coast Guard cutter *Mackinaw*, which was anchored to serve as an orientation point. Around us were hundreds of boats—273 of them registered for the race, plus many powerboats and motor yachts filled with spectators. The city rose in tiers to the sky, skewing perspective. Was it a mile away or ten?

We were in open water now, exposed to the biggest waves. They were typical of the Great Lakes—not rollers, but steep, short-period wind waves. Freshwater is less dense than salt water, so lake waves rise quicker and run faster and can be harder for a boat to negotiate than the long rollers of saltwater seas. Now they rushed headlong down the length of Lake Michigan and slammed into the wakes of the powerboats around us, making a confused chop. At forty-four feet long, *Gauntlet* handled the turbulence better than the smaller boats could, but we were thrown around enough to be uncomfortable.

Everywhere we looked, sailboats heeled in the wind or veered suddenly away from one another. "Starboard! Starboard!" crewmen yelled if they had the right of way and others were slow to yield. According to the Rules of the Road, a boat that is tacking or jibing must give way to one that is not, and a boat on a port tack must give way to one on a starboard tack. If they're on the same tack, the boat to windward must give way to the one at leeward. But not everyone seemed familiar with the Rules of the Road.

Sailboat races are organized within two broad categories. One-design races are for boats of the same size and type and thus with the same potential speed. Handicap races—the Macs fit this category—allow boats of many kinds to compete equally, using formulas that factor length, weight, hull style, and sail dimensions. Entries are grouped within several divisions and sections. We were in Section 2 of the Performance Handicap Racing Federation Division, placing us in a category perhaps best described as "Serious Amateurs," although by definition everyone in the race was an amateur. As in most sailboat regattas, we competed not for cash but bragging rights. Each divisional winner gets his name engraved on a trophy displayed in the clubhouse, and winners can fly a champion's pennant for a year. If a boat is extraordinarily successful—if it sets a course record or wins repeatedly—its photo is hung in the clubhouse.

The slowest boats began the race at noon. Other sections were staggered at fifteen-minute intervals. Ours, with twenty-three boats measuring from thirty-five to sixty-four feet in length, would leave at two-fifteen. Last to start, at three o'clock, were a baker's dozen of "big 70s," the largest and fastest in the fleet.

Guy asked me to sit near him by the helm and take charge of trimming the mainsail. It was a matter of cranking a winch to sheet in the main sheet or release it to ease the sheet out. After a few tries I felt I had it mastered.

At our appointed time all the boats in our division got a running start and sprinted together across the starting line at the committee boat. With her sails close-hauled into the wind, *Gauntlet* climbed each wave and crashed into each trough. Right out of the gate we were passing other boats. I trimmed the mainsail at Guy's command. Sheet in, ease out. Piece of cake.

Or maybe not. Even in the harbor the pitching of the boat and the smell of diesel exhaust wafting over the water had bothered me. Lack of sleep didn't help. Neither did my performance anxiety. Nor the lumberjack breakfast I'd packed in. I had been seasick a few times in my life, and was determined not to let it happen now. Willpower would carry me through. Besides, I had taken the recommended dosage of Dramamine. And I was wearing wristbands—those pressure-point elastics that seem more cabala than science, except that thousands of people swear they're effective. I would not get sick. I refused to get sick. It was important that I—who had bragged about being around water all his life, about not being deadweight, about being a fast learner and a hard worker and an all-around great guy—not get sick.

I got sick.

Bill Craig and Scott Jacobson timed me. They were busy, but not too busy to make a note for reference: One hour and three minutes from the start of the race. I think a betting pool was involved.

The boat rose on every wave and crashed into every trough, and I vomited every thirty minutes for the rest of the day. In some of the intervals I felt better, even well enough at times to return to my post at the mainsail winch, though it was obvious that I wasn't needed. Guy said I should keep my eyes on the horizon, so I stared astern and watched Chicago, a cluster of hazy skyscrapers rising beside the lake like the Emerald City of Oz. Each time I looked, the skyscrapers were a bit lower in the water. Then I would lie on the deck and die a little. The pattern kept repeating: cold sweats, roiling stomach, a flood of saliva—then retching, expulsion, humiliation.

The waves topped out at sixes and eights, so big that Guy had to steer at an angle down the larger crests because if he took them straight on, the boat slammed into the troughs with such force that the aluminum deck flexed a few inches. I knew the others were concerned about this, but it was of no importance to me.

Somebody asked if I felt like I was going to die.

I managed to raise my head. "Yes," I said.

"Good. That's the next-to-last stage."

"What's the last stage?"

"You *wish* you would die."

An old joke. Perhaps it was funny once.

Hanging with my face inches from the hissing lake, having just expelled a meal I remembered eating in my brother's apartment in New York in 1985, I looked up to see one of the big 70s passing close on our starboard, bound for glory. A dozen guys in red foul-weather gear sat along the rail watching me. Most of them were grinning.

In *The Varieties of Religious Experience*, William James classified *mal de mer* as a mild state of depression he called "anhedonia," the symptoms of which are "passive joylessness and dreariness, discouragement, dejection, lack of taste and zest and spring."

The esteemed professor was never seasick. He couldn't have been. The condition he describes would be remedied by a hot bath and a nap. Nothing about my joylessness was passive. My joylessness was extremely active.

Late in the day, as the sun went low, the waves diminished to fours and sixes. Slowly, incrementally, I began feeling better. I tried a sip of water. Then I tried half a stick of ginger-flavored chewing gum. Then more water and a single saltine cracker.

I noticed the lake again, blue and oceanlike. The Michigan shore showed low at the limits of sight, with a line of cumulus clouds bunched above it. Around us whitecaps rode the waves—whiter whitecaps than I had ever seen—and on every side and at every distance were sails, colored mostly white but also pink, orange, yellow, and red. Everywhere was a great deal of blue water, topped with a half-dozen shades of sky fading to the horizon. It was glorious. Who had time to be sick?

I climbed to my feet.

"Okay. Put me to work. I want to learn the ropes."

"Look who's back from the dead."

Guy sat relaxed at the helm. This was his twentieth Mac. He's a CPA by profession, owner of a firm in Grand Rapids employing a dozen accountants and with a client list that includes a number of NFL players.

I apologized for getting sick.

"Hey, don't feel bad," he said. "I've seen three–hundred–pound line-backers turn to jelly as soon as they hit rough water. There's no shame in being a pussy."

For the first time I participated as we tacked. With the mainsail and jib close-hauled, Guy called: "Ready about!" and everyone cleared the sheets and got ready to move. "Hard alee!" Guy said, steering the boat straight into the wind, and we slid inboard and let the jib and main sheets fly. The mainsail went slack in great billowing luffs, and we ducked beneath the boom as the bow of the boat passed through the wind. We jumped to the winches and began grinding as rapidly as possible to sheet in the main and jib sheets. The boom swung and the mainsail filled with a snap so powerful I could feel it from my feet to my scalp. The boat surged into the wind.

I asked Guy about *Gauntlet*. She was a Kauffman 44-footer, built in Italy in 1975, aluminum hull and deck, Marconi rig. In 1979, manned by an Italian team, she survived an infamous Force 10 gale that struck the Fastnet Race in the Irish Sea, capsizing seventy-seven boats, knocking down a hundred others, and drowning fifteen sailors. She was a barebones racer. No frills, no extra weight. Below deck was naked superstructure, with a few plywood berths barely big enough to lie on. No cabins, no galley, no privacy. Her chemical toilet sat isolated in the open below deck, a humble throne surrounded by stacks of sails.

Bill Craig joined us in the stern. Like most of the others, he was from the Grand Rapids area. He was a forty-seven-year-old banker who had sailed in a dozen Port Huron–to–Mackinac races but only a few Chi-Macs. He preferred the Port Huron Race, he said, because he had grown up sailing it with his late father, Rocky.

Now, for my benefit, Bill discussed general principles. "Sailing sucks," he said. "The same aerodynamic law that causes lift beneath an airplane's wing creates pull on a sail. The wind doesn't push, it *sucks* the boat forward."

"What amazes me," Guy added, "is that a sailboat uses one fluid substance—wind—to overcome the resistance of another fluid substance—water. To me, it seems miraculous. Water clings. It's displaced for a few moments as the boat cuts through, but then returns to its former place. It's a wonder we can move at all."

Water swept past the hull. It gave the illusion of speed, though I knew we were making less than ten miles an hour. Still, the boat rose eagerly on every wave and seemed to leap from its crest.

"She's a dog with a bone in her teeth now," Jeff Burt said. Jeff was thirty-six, a mortgage loan officer, and had sailed since he was six years old. He had been in two previous Macs, but this was his first aboard *Gauntlet*. As a boat owner himself, he was accustomed to running things and would distinguish himself as an excellent sailor and probably the most reckless member of the crew. He said his nickname as a kid should have been "Runs with Scissors."

Chuck Kmiec came up from below and sat in the cockpit. He was navigator, cook, and resident Mensa genius, fifty years old, an old friend of Guy's, and the big man on board—not three hundred pounds, maybe, but crowding it. Though he'd been racing for two dozen years and in seventeen Macs, he grew up a troubled kid in inner-city Chicago, far from the sailing world. When he was twelve, his mother dragged him to a Boy Scouts meeting. He liked it. He excelled at scouting and liked the feeling so much that he went on to excel in academics as well. Now he had a string of university degrees and worked as a computer consultant and ran a scout troop of his own near his old neighborhood. "Sailing is the only sport that uses all my scout skills," Chuck said. "I like the resourcefulness it demands, and how tough it can be. All other sports are wimpy."

The rest of the crew sat at the rails looking over the water. They ranged in age from sixteen-year-old Zach Schramm, a scouting protégé of Chuck's, to fifty-eight-year-old John "B.B." Whitton, a veteran of dozens of races who though bitter to discover that I was deadweight after all was getting over it. Occupations varied widely. Doug Van Der Aa was a lawyer and CPA. Todd Suess worked as a boiler operator at a factory in Grand Rapids. The single constant the crew had in common was a passion for sailing. Together, they had tallied more than a hundred Mac races.

Gradually the wind swung from north to west and finally to southwest. By nightfall we had made two headsail changes and tried three or four spinnakers, none of which was quite ideal. The crew tinkered constantly with the sails, trying to milk any advantage from the wind. A slight correction will sometimes mean an extra knot or two. At an average speed of seven knots (about eight miles per hour) a boat can complete the race to Mackinac in forty-two hours. Add a knot and you finish in thirty-seven hours. It can be the difference between first and last in your division.

With night approaching, the wind fell and our progress slowed. Stars showed first in the eastern sky, then in the west. Mastlights twinkled in every direction around us—but more of them astern than ahead.

By then we were in the middle of the lake, no land visible in any direction. Chicago lay far behind, but the sky over it glowed as if a fire burned beyond the horizon. To the west the sky was lit in a dome above Milwaukee; to the east it glowed above Grand Haven and Muskegon. Most of the way up the lake we would see the residual lights of cities, each a radiant stain on the horizon, growing smaller as the cities grew smaller to the north.

Bill and I sat at the rail with our legs swinging free over the water. He told me that after his father's death in 1997, he wrapped his ashes in small bundles of paper towel and took them along on the Port Huron Race, which his father had enjoyed more than anything else he did, and dropped them overboard at key places along the route. The last bundle he saved for the drive home and threw over the side of the Mackinac Bridge.

Now he pointed out constellations. He knew them well, from lessons learned long ago from Rocky, who loved the night sky almost as much as he loved sailing. We looked up in silence at the stars—and at that moment a meteor flared across half the sky and left a fading trail.

Late in the night I went below and slept for a few hours.

Morning dawned bright and cool, with a fresh wind from the west and waves three to five feet. Sails dotted every horizon.

Guy gave me new assignments. He showed me how to tighten the boom vang to keep the boom horizontal. He instructed me to watch the luff of the mainsail, easing or trimming as required to keep it taut. Ease too much and the sail "bubbled" with slack. Trim too much and the boat heeled over.

I was starting to learn how much there was to learn. Sailing requires technical mastery of rigging and navigation, knowledge of water and weather, a feel for the boat. A team mentality is at work—as much, or more, as on any sports team—which explains why it is so difficult for a novice to enter a race. Nobody wants to introduce an unknown to the mix, let alone one without experience. A crew must blend together well and efficiently. The more quickly every sail is changed, the more smoothly every tack and jibe is performed, the faster the vessel goes. If one member screws up, the system collapses. When everyone does his job right, a change of sail or tack works with precision, like a well-executed football play.

I discovered the pleasure of "rolling" the competition. We would close on a boat and pass on the windward side, stealing the air, forcing the other crew to tack across our stern, where they rolled in our wake. If we were feeling cocky we crossed their bow, an act of high disdain certain to piss them off.

It was fun to pass boats larger than ours or those in faster divisions—smaller boats were "ratboats" unworthy of our attention—but we reserved our greatest zeal for the others in our division. We sparred frequently with *Blind Hog, C. C. Rider, Dr. Detroit*. Feelings ran strong against *Vagary*, which once set a divisional record but was disqualified for not having a spare tiller on board. One of our crew, twenty-six-year-old Scott Jacobson, had sailed for several years on *Vagary*, but only recreationally. "I wasn't a rock star yet," he said, "so they wouldn't let me race."

"What's that ratboat?" Scott asked, glassing a boat ahead of us. It was Bob's, the one I was originally supposed to be on. I had already told the story, and everyone was indignant on my behalf. Guy cut upwind and very close and we drew abreast of the other boat, robbing her air and making her sails go slack and her rigging clank against the mast. Bob and his crew looked none too frisky. Their boat was eleven feet smaller than ours, and the waves had been tough on them. We cut across their bow, a flagrant diss. I stood in the stern and waved, pretty sure that Bob recognized me. He raised his middle finger halfheartedly, but his eyes went down in submission.

Early the third morning I climbed to the deck and discovered a world obscured by whiteness. Jeff and Scott were on duty alone. They'd been up all night and were nearly comatose. The sails hung limp, and the boat sat motionless at the center of a thirty-foot circle of flat gray water. Lake Michigan was out there, but invisible. We could see nothing around us but fog.

"A CSS day," Jeff said.

"What's that?"

"Can't see shit."

On the navigation chart I saw that we were a few miles off Frankfort, approaching Point Betsie, with Sleeping Bear Dunes and the Manitou Passage ahead—my stomping grounds, and the most picturesque leg of the race. Henry Rowe Schoolcraft paddled along this shore in September 1820 during a five-month canoe expedition around the Great Lakes led by Michigan's first governor, Lewis Cass. Schoolcraft was a geologist, ethnologist, and Indian agent who is best remembered for discovering the source of the Mississippi in 1832 and for his report on the legends of the native tribes of Superior, *Algic Researches* (1839), which inspired Longfellow's epic poem, *Hiawatha*. Schoolcraft's journals offer an interesting perspective on the lakes and the people who lived around them and are a reliable record of many details of physical geography. Observing the Lake Michigan shore from Point Betsie to the Manitou Passage, Schoolcraft wrote: "There is a great uniformity in the appearance of the coast, which is characterized by sand banks, and pines . . . the beech and maple are occasionally intermixed with the predominating pines of the forest." He found greater variety at Sleeping Bear Dunes: "The shore of the lake here, consists of a bank of sand, probably two hundred feet high, and extending eight or nine miles, without any vegetation, except a small hillock, about the centre, which is covered with pines and poplars, and has served to give name to the place, from a rude resemblance it has, when viewed at a distance, to a couchant bear."

A century earlier, in August 1721, the Jesuit missionary Pierre de Charlevoix came this way also, and commented in a similar vein: "I perceived on a sandy eminence a kind of grove or thicket, which when you are abreast of it, has the figure of an animal lying down: the French call this the Sleeping [Bear] and the Indians the Crouching Bear."

In the Ojibwa legend of Sleeping Bear, the dunes were formed by a sow bear that collapsed exhausted on the beach after swimming across Lake Michigan to escape a Wisconsin forest fire. Behind her, unable to reach shore, her twin cubs faltered and sank beneath the waves, forming South and North Manitou islands.

Sleeping Bear Dunes haven't changed much since the days of Charlevoix and Schoolcraft, though at 460 feet they're twice as high as Schoolcraft thought. The "bear" is still visible at the top, but erosion has shrunk the hillock at the center and the grove has been reduced to a few gnarled trees.

Dunes are found in many places around the Great Lakes, but never in concentrations such as those along the eastern shore of Lake Michigan.

Ranging from a dozen feet to several hundred feet high, extending from the edge of the lake to a mile or more inland, they form the most extensive network of freshwater dunes in the world. Carl Sandburg declared (with some exaggeration) that "they are to the Midwest what the Grand Canyon is to Arizona and the Yosemite to California. They constitute a signature of time and eternity."

That signature has turned out to be indelible in only a few places. One is at Sleeping Bear, where the dunes are contained within a national lakeshore that has protected them from bulldozers and front-end loaders. Other dunes to the south are similiarly protected within Michigan state parks near Manistee, Ludington, Muskegon, Holland, South Haven, Benton Harbor, and elsewhere.

At the extreme southern end of the lake, portions of 14,000-acre Indiana Dunes National Lakeshore were set aside starting in 1927, but not before most of that shoreline had been leveled and built over with houses, factories, and power plants. The remaining dunes are squeezed between the industrial metropolises of Michigan City and Gary, just down the shore from Chicago, and comprise the most fragmented national park in America. From the lake, looking shoreward, the dunes are overwhelmed by steel mills and other structures. But up close, and especially back from the shore, they form an ecosystem remarkable for its stark beauty and the amazing diversity of plants and animals it supports.

In the late nineteenth century, a young professor of biology named Henry Chandler Cowles approached the Indiana Dunes as if it were a living laboratory. While studying the pioneering of the land by plant communities, he observed that dunes evolved from barren sand near the shore to ridges of pioneer grasses to hills of shrubs and trees and finally to climax forests. The plants that lived on the sand, he discovered, grew in predictable patterns, with marram and sand reed grasses first, followed by red osier dogwood and sand cherry and cottonwood, then maple, oak, and pine.

Others before Henry Cowles had recognized that the Indiana Dunes were a dynamic ecosystem, with land forms and microclimates supporting more plant diversity per acre than in any other national park in the United

States. But where others had seen only hills of sand and an interesting variety of plants, Cowles saw centuries of ecological progress compressed into distinct zones only a few hundred feet apart. In 1899, when he published his observations in a report, *The Ecological Relationships of the Vegetations of the Sand Dunes of Lake Michigan,* it sent a shock wave through the scientific world. Cowles had demonstrated for the first time that plant communities succeed one another, each serving as the foundation for those to come, while simultaneously creating the conditions for its own collapse. This concept of the interrelationship of organisms was revolutionary and it changed the way people looked at the natural world. Some historians now mark Cowles's paper as the beginning of the science of ecology.

Though parks have thrown a protective barrier around many Lake Michigan dunes, hundreds of miles of others are privately owned and have succumbed to commercial and residential development. Homeowners can't be blamed for wanting to live along the lakeshore, in spite of history's lesson that it is unwise to build a house on sand. During the 1970s, when water levels were the highest in historical times, hundreds of lakeside houses and cottages washed into the lake. Undeterred, developers continued to bulldoze the tops from dunes to unblock the view of the water and have leveled countless others as building sites for houses, condominiums, and town houses. Even where development is restricted, dunes are easily damaged by the trampling feet of sunbathers and the wheels of all-terrain vehicles.

Dunes can form in many places. All that's required are sand, wind, a few obstructions, and time. Sand is the product of force applied to rock. Around the Great Lakes, most of the force was originally supplied by millions of years of glaciers grinding over bedrock. The rock was crushed into fragments, which in turn were ground to grains of sand. About ninety percent of the grains were made of quartz, and less than ten percent were feldspar. Some were magnetite, a very fine ironlike mineral that accumulates in thin black layers often mistaken for oil or other contaminants. The rest were a mix of garnet, calcite, ilmenite, hornblende, and epidote.

As the glaciers receded, rivers of meltwater washed through the debris they left behind, carrying sand downstream to the basins that would become the Great Lakes. As the lakes formed, current and waves pushed the sand shoreward, forming beaches. When the sand dried, every grain became subject to wind.

It takes a wind of about eight miles per hour to make fine sand fly, and about twenty-five miles per hour to lift coarse sand. Kneel on a beach on a windy day and you can see a mist of sand blowing a few inches to a few feet above the ground. Look closer and you can watch individual grains jumping and tumbling along. They go airborne, collide with one another, fall suddenly into eddies behind rocks and other obstructions. This bouncing transport is called saltation. During lesser winds, the grains roll along the beach in a process called surface creep. Over time, collisions chip away at the grains, rounding and polishing them. The older the grains, the rounder and smoother and more uniform in size they become.

Sand would migrate forever inland if not for marram grass. Also called American beachgrass, and belonging to a group of about a hundred similar species found along the Great Lakes, Atlantic, and Pacific coasts, marram is among the first plants to grow on a beach and serves an important function as a dune builder. When a clump of the grass takes root, it creates a slight break in the wind, allowing grains of sand to fall behind it. As the sand collects into a pile, more grains fall in its eddy, building a ridge with a long slope on the windward side and a steep slope on the lee side—the classic shape of a dune. The rising sand buries the plant, which sends out underground stems called rhizomes that shoot new sprouts to the surface. The network of rhizomes and roots bind the growing dune in place. As the dune ages, organic material accumulates on its surface and other plants take root and gradually crowd out the marram. In time, a hill of sand covered with a thin layer of loam will become a forest of mature hardwoods and conifers.

A variety of dunes is found around the lakes. Sleeping Bear is a "perched dune"—a crest of sand sitting on a glacial moraine. The winds at Sleeping Bear Point are strong enough to blow sand up the face of the moraine, depositing it at the top. Grand Sable Dunes on Lake Superior were formed in a similar way.

The most common dunes along the lakes are called linear dunes, and form in ridges parallel to the shore. Rarely higher than about fifty feet, they rise row after row away from the water's edge, becoming wooded ridges as they age. From the air or from a boat far out on the water they look like tightly fitted wrinkles on the land.

Parabolic dunes are U-shaped and occur when plants that stabilize a linear dune are destroyed. Once the plants are gone—killed by natural processes or from human activities—the sand becomes exposed to wind that blows it into a saddle-shaped ridge of bare sand known as a blowout. Sometimes one blowout will stack upon others until they form a giant parabolic dune several hundred feet high. Other blowouts become "walking dunes" that travel inland a few feet a year and bury everything in their way, including houses. On the shore of Lake Michigan in the 1800s, an entire abandoned fishing village was buried this way.

Since about 1920, the sand from Lake Michigan dunes has been strip-mined and sold for industrial use. About ninety-five percent of it goes to foundries that cast engine blocks and other parts for the automobile industry, while smaller amounts are used to make glass, sandpaper, toothpaste, and other products. In a foundry, the sand is combined with a binding material and shaped into molds, into which is poured molten metal. Once the metal cools and hardens, the sand is broken away, leaving the finished product. About four hundred pounds of sand are required to cast an engine block in this way.

Until 1976, sand mining in Michigan was unregulated; mining companies bought shoreline dunes wherever they could and stripped away the sand as fast as they could sell it. One of the largest barrier dunes on the Lake Michigan shore, Pigeon Hill in Muskegon, originally measured about forty acres in area at its base and rose to a height of two to three hundred feet. Sand Products Company began excavating it in 1936. By the mid-1960s, Pigeon Hill was gone, replaced by a hole in the ground.

Aesthetics are not the only reason to protect dunes. Many of the plants and animals that live among the dunes can survive no place else. A number of them are endangered or threatened, including the piping plover, a rare shorebird that nests near the water's edge in northern Lake Michigan, and Pitcher's thistle, a flowering plant found in isolated patches among the dunes.

Michigan's Sand Dune Protection and Management Act of 1976 saved some of the more critical dune habitats from mining, but environmentalists have recently raised questions about its effectiveness. The general public and environmental groups took for granted that the act protected existing dunes and would result in a gradual phasing out of sand mining. This seemed inevitable after Ford Motor Company adopted a policy of not using sand from barrier dunes in its foundries, relying instead on sand mined from interior lands that were less ecologically sensitive.

In fact, however, sand mining continues at only a slightly slower pace than before regulation. In 1976, fifteen active mining sites were in operation on a total of 3,228 acres of dunes; twenty-five years later, there are twenty active sites, totaling 4,848 acres. The output of sand is slightly lower than it was in 1976, but still averages about 2.5 million tons per year. In all, some 2.3 million dump-truck loads have been hauled away since 1976. And they've been sold cheaply: the going rate for dunes is five to ten dollars per ton.

From the *Gauntlet* that morning we watched the sun rise slowly, an orange glow in the whiteness, as the fog disintegrated a droplet at a time. In the distance a few gauzelike images of boats appeared. Ghost ships. Then the fog lifted altogether, and ahead of us was open water to the horizon with not a single sail to be seen. We turned to look behind and the lake was filled with them.

Scott leaped to his feet and hooted. "We're gonna roll in ahead of the seventies!"

Jeff fiddled with the FM radio, lifting it and turning it, trying for a clear station. Then he found a classic rock station playing Led Zeppelin's "Whole Lotta Love." He cranked the volume as high as it would go, and the morning cracked open and electric guitars spilled out. We raised our fists in the air and tipped our heads back and howled. The rest of the crew poured topside, awakened by the tumult. As if by decree, a breeze came up from the southwest, riffles streaked across the water toward us, the sails caught and filled, and the boat jumped to life beneath us. We banged the corner around Point Betsie and headed toward the Manitous. The wind increased.

"We're getting aggressive now," Jeff said.

It was perfect, nothing in the world but water and sun and blue sky and a good wind from the southwest at fifteen knots. We entered the Manitou Passage, where dozens of ships have been lost, often in fog, more often in storms. The passage cuts inside the islands, with Sleeping Bear Dunes dominating the mainland—a khaki tsunami of sand—and with South Manitou Island off our port side. A hundred yards or so off the island's south tip sat the rusting and guano-streaked hull of the *Francisco Morazan*, a Liberian freighter that grounded there during a storm in November 1960.

The winds and currents in the Manitou Passage are often treacherous. Schoolcraft, in his *Narrative Journal of Travels*, mentions in a footnote chilling in its offhandedness that fifty Indian canoes were once lost here, somewhere between the Manitou Islands and Sleeping Bear Dunes. The details of what happened have been erased by time. Perhaps a war party of fifty canoes was on its way to ambush an enemy camp in Wisconsin. Or an entire village population of men, women, and children were paddling to the islands to make maple sugar or pick blueberries or rendezvous with friends and relatives. Probably they were caught in a squall. They come up quickly here, and gain force as they funnel between the islands and the mainland. Fifty canoes, each carrying at least two and probably three or more people. Gone without a trace.

In 1838, the French naturalist Francis Comte de Castlenau encountered a storm while sailing through the same passage. He later wrote: ". . . we were a plaything of the giant waves that pushed us toward the immense bank of sand [Sleeping Bear Dunes] . . . I have seen the storms of the Channel, those of the Ocean, the squalls off the banks of Newfoundland, those on the coasts of America, and the hurricanes of the Gulf of Mexico. Nowhere have I witnessed the fury of the elements comparable to that found on this fresh water sea."

We passed the "Crib"—an automated lighthouse at North Manitou Shoal. We passed Pyramid Point, Whaleback Mountain, Sugar Loaf—all prominent remnants of glacial plowing on the mainland. We could see the tracks the ice made there, and the level bench twenty feet high where the shore of the lake once stood. Good Harbor Bay was a long, low strip of sand backed by green hills, with houses visible along the backdunes near Leland. It was where I would spend six weeks in February and March, getting acquainted with wind, waves, and sand.

Guy canted the boat to fifteen degrees and we jumped a knot in speed. "Our waterline is starting to pay off," he said. I looked back and every sail on the horizon was canted at the same angle.

Eight knots now, and it was impossible not to be mesmerized by the sensations of sailing—the wind in our faces, the slapping bow waves and creaking sails, the impish sense that we had reached up and snagged the tailhook of the sky. We had caught a free ride across the planet.

We closed on some 40s and 50s that had been so far ahead in the morning we couldn't see them. We caught up to *Captain Blood* and passed her. Beyond North Manitou, we passed *Blind Hog* and *Kokomo* and *Flying Cloud*. It was pure motion. It was perpetual. It was better than rock-'n'-roll.

Then we fell into a wind hole. Two hazards of sailing are too much wind and too little wind. On the Great Lakes, they can sometimes be encountered within moments of each other. For twenty minutes we watched boats we had passed catch up. Some steered close to shore, hoping to find land breezes. Others drifted to a halt a hundred yards from us. The water was glass-calm. Here and there streaks showed on the surface, like the patterns you can make walking across freshly vacuumed carpet. A stirring would crawl northward, a hundred feet wide and a half-mile long, crossing the water at the speed of cloud shadow, and if we were lucky it intercepted our path and snapped our sails full and we jumped to four or five knots and passed a few boats before it got away and we settled again on glass. A couple boats farther out seized a wind and rode and rode and rode; we were jealous. Then we seized one ourselves and were off and passing South Fox Island, with the mainland slipping lower on the horizon.

We passed Grand Traverse Bay, which at this distance looked narrow, like the mouth of a bottle, though I knew it was ten miles wide. My home was down there, a few miles past the curve of the earth.

We passed North Fox Island. We passed Little Traverse Bay and Beaver Island, low and green. Now the wind was strong and steady, with no gaps, and the race was heating up.

By sunset, we had abandoned all sense of leisure aboard *Gauntlet*. Sailing was serious business now. Northern Lake Michigan is colder and wilder and less forgiving than the rest of the lake. The bottom there is studded with dolomite shoals that can gut a boat. Ahead was Gray's Reef, legendary trouble spot. Beyond it we would bear east into the Straits of Mackinac, enter Lake Huron beneath the Mackinac Bridge, and cross the finish line at Mackinac Island.

We approached Gray's Reef in darkness, clouds blacking the sky, nothing to see but buoy lights and mastlights. The wind was strong and dead astern, and we ran ahead of it, our big spinnaker up and mainsail out and everyone standing by at stations. We crowded a pod of boats, while others challenged us, staying hard on our beams and stern. But we were passing more often than not.

Coming into the channel at Gray's Reef, with red and green buoys on either side, we made a move on a boat ahead of us. We cut to the inside and laughed when her crew shouted through the darkness. The boats chasing us fell behind.

It was a sprint to the finish now, everyone throwing everything they had before the wind. With White Shoal Lighthouse off to our port, we spun to starboard on the last green buoy, our rail inches from it. We were working without words, on automatic, ten men operating like a single organism. We raised the jib as we dropped the spinnaker, not losing momentum for even a moment, hauling the house-size tent of nylon into our arms as fast as it fell, never letting it touch water, shouting with the elation of it. The boat caught the beam wind full in the jib and mainsail, and leaped like she was spurred.

Damn. We *pounced* into that wind. Behind us was gridlock, chaos flapping at the turn. We heard hulls bumping hulls and madmen yelling. We heard "Starboard! Starboard! Starboard!" and the word "fuck" manifested in a dozen ways. Mastlights swayed. Rigging clanked.

Now we were cutting hot. Coming into the Straits with the five-mile-long bridge spanning it in an arc of lights, like rockets sprayed from land to land, we made 8 and then 8.2 knots, but with the wind in our faces it felt like 20. It felt like 60.

We passed beneath the bridge cheering. We figured we were first or second in our division. Just five miles ahead was the island. We could see the long antebellum front porch of the Grand Hotel lit in welcome. Parties were in progress all over the island. All we had to do was pass the buoy boat at the entrance to the harbor and somebody would touch off a cannon, hit our sails with a spotlight to identify us, and enter our time in the books. We were ahead of many of the 50s and some of the 60s. Nothing stood between us and glory.

And the wind died.

It died a sudden death. Went out like a candle.

For an hour and a half, we bobbed. Boats drifted up and bobbed with us. Every ratboat we'd passed in the Manitou Passage, every barge we'd blasted out of the water at Gray's Reef—caught up to us. We became just another vessel in a drifting armada. Some boats had enough momentum to pass us. A phantom puff would goose them and they would glide a hundred yards and slow and stop and sit there. I wanted to huff and puff and blow our sails full. But all we could do was wait.

Finally, a breeze mussed the water and we edged forward. At 4:25 A.M. we crawled across the finish line, moving too slowly to register on our GPS. A spotlight caught our rigging, our neon numbers glowed, a cannon banged. But it was only a formality. No longer did it matter who finished first. Everyone was a winner. Time to celebrate.

But at that hour most of the edge was lost from the party. In the harbor, we tied up to boats already rafted several deep and popped a bottle of champagne. We passed it around. We gripped each other by the hands and congratulated ourselves. A few revelers nearby gave some hoots and hollers, but you could tell their hearts weren't in it.

We had set out from Chicago with our destination clear, and some of us stayed on course and some of us got lost, and some performed magnificently and others not well at all. We knew moments of elation, discontent, triumph, and discouragement. We suffered a little. We learned a little. We threw ourselves at the mercy of forces we had no control over while managing, barely, to harness lesser forces. Then the race was over and we had used up all our strength. We had pushed our limits, had tried our best, had seen how we measured up (not as well as we'd hoped), and now we were weary of the whole business.

In the gray dawn I went ashore and walked into town. The ground kept tipping away from me, its fixity an illusion. No motorized vehicles are allowed on Mackinac Island—transportation is by foot, bike, or horse—and the streets gleamed where a guy with a firehose sprayed away yesterday's horse droppings. Sailors walked alone or in small groups, their faces stunned with exhaustion.

The Pink Pony is supposed to be where you go after the race to get drunk and hit on each other's wives and girlfriends. I stuck my head inside. It was a nice place, with lots of oak and brass, and crowded with unshaven, sunburned men standing around with drinks in their hands. One or two were shouting lustily.

I might have fit right in. When we spun *Gauntlet* around the last buoy at Gray's Reef and dropped the spinnaker, all of us drunk with speed and conquest, Jeff or Scott or someone had slapped me on the back and shouted that I was a rock star now. I didn't believe it, but I liked the sound of it. Here I was, filthy, sun-cooked, hardly any sleep for four days, hands raw from rope damage—the Pink Pony was just the place for a rock star like me.

But, no. Not this time.

Instead, I called home, caught the ferry to the mainland, and sat on a curb in the parking lot to wait for my wife. An hour and a half later she pulled up in our truck and lowered the window. She gave me a funny grin.

"Hey Sailor," she said. "Going my way?"

Chapter 5

THE STRAITS OF MACKINAC

Mighty Mac ◆ The Crossroads of the Lakes ◆ Wind and Ice ◆
A Brief History of the Straits

Ten months later I stood in moonlight on the deck of the *Malabar* and watched the Mackinac Bridge pass overhead. From a distance it had seemed grand—a curve of pure genius, radiant and dazzling—but from beneath it was merely grandiose. Its steel underbelly looked old and frail, nothing like an engineering feat for the ages. Only Hajo and I were on deck. Matt, Harold, and Tim had asked to be awakened at the bridge, but we decided not to disturb them.

Up here in Michigan we sometimes call the bridge "Mighty Mac" or "Big Mac," though the latter fell out of favor after the burger of the same name became so famous. Usually we just call it "The Bridge." We say, "The wind was so strong traffic was stopped at The Bridge," or, "My vacation never starts until I'm over The Bridge." If you want to cross between the peninsulas of Michigan, the only way to avoid The Bridge is to make long detours south around Lake Michigan or Lake Huron and come up through Wisconsin or Ontario. But there's no reason to avoid it. The drive across takes ten minutes, and traffic is heavy only on holiday weekends. Halfway across you're two hundred feet above the water, with a splendid view of the Straits—Lake Michigan on one side, Lake Huron on the other, with unmarred horizons both ways. As long as I can remember, the toll for automobiles has been a buck and a half. A bargain. But pedestrians and cyclists are prohibited. You can cross under your own power only during the annual Labor Day walk led by the governor.

(Though once, when I was a kid, my brother and a couple of friends and I set out to walk it. Our families had made camp for the weekend at an old municipal campground—gone now—on the shore in Mackinaw City, and we boys had gotten up earlier than our parents and gone exploring. We boosted each other up to a service ladder at the first concrete pier of the bridge and climbed to the beams and bracings under the roadway. We found a steel grid catwalk there, and followed it a mile or so into the Straits, intending, I think, to cross to the Upper Peninsula. I remember the catwalk disappeared in a perspective point far ahead of us, out where the center span of the bridge arched to its highest elevation, and that it exerted an irresistible pull. As we walked, traffic roared a few feet over our heads, and the waves in the Straits grew smaller beneath us and the wind grew stronger. At some point the catwalk began trembling and we looked back to see a security officer jogging heavily after us. We waited for him to catch up. He was not a happy man.)

All the Mackinacs, incidentally—bridge, Straits, fort, and island—are pronounced "MACK-in-awe." Visitors find this confusing. Only a few variations of the name, such as the town of Mackinaw City and the Coast Guard cutter *Mackinaw,* are spelled the way they sound.

When it was opened in 1957, the Mackinac Bridge was the longest suspension bridge in the world. The suspended portion measured 8,614 feet between anchorages, while the steel superstructures and their approaches added almost 18,000 feet, for a total length of 26,444 feet, or just over five miles. Construction took three and a half years and required an armada of marine construction equipment that at the time was the largest ever gathered. The designer, David B. Steinman, and the hundreds of engineers and workers who translated his blueprints into cement and steel faced formidable challenges. For more than a century, even the staunchest advocates of a Straits bridge had wondered if it was possible to build one there. Critics said the bedrock on the bottom was too weak to support such an enormous structure; that currents were too strong; that the winds that funneled through the Straits would topple any bridge; that ice in the winter would demolish the stoutest concrete piers. Steinman and his army of workers accomplished what many had thought was impossible.

Before the bridge and before the I-75 expressway leading to it, when the only highways in northern Michigan were two-laners notorious for their potholes and frost heaves, small boats ferried passengers and freight in summer, and sledges hauled them over the ice in winter. One particularly cold winter around the turn of the twentieth century, a railroad track was laid across the ice and trains rumbled over the Straits. Later, ferries carried boxcars and automobiles across. Though traffic generally was light in the tranquil 1940s and 1950s, the bottleneck at the ferry dock could be daunting. In mid-November, at the opening of deer season, thousands of hunters from around the Midwest would converge on Mackinaw City and wait for their turn to cross to the Upper Peninsula (or "U.P.," as it's known). A line of traffic at the ferry dock backed up fifteen or twenty miles. Drivers sometimes had to wait all day and night to get a spot on the ferry. If the weather acted up, as it often does in November, the hunters had to wait even longer.

Not everyone objected to those long waits. The late John Voelker, native of the U.P. (and thus a "Yooper"), former Michigan Supreme Court justice, and the author, under the pseudonym Robert Traver, of *Anatomy of a Murder* and ten other books, was outspoken in his wish to keep the U.P. isolated from the more crowded Lower Peninsula. Michigan governor G. Mennen Williams invited Voelker to be among the dignitaries present at the opening of the Mackinac Bridge, but he declined, citing a conflict of interest. He told the governor he was already founder and president of the "Bomb the Bridge Committee."

The U.P. today is still among the least populated regions of the lower forty-eight states. It's a rough-and-tumble land of woods and swamps and four-corner towns, where logging remains the only industry to thrive (iron mining was once the mainstay, but 598 of the peninsula's 600 mines have closed, victims of the ailing domestic steel industry), and the largest city, Marquette, has a population of less than twenty-five thousand. But Voelker had reason to be worried. Strings of franchise businesses, busy highways, crowded campgrounds, a real estate boom that parcels off thousands of acres of forest to sell as vacation property—all are the legacy of The Bridge.

The *Malabar* pushed steadily eastward. Hajo had long ago adjusted the speed of the boat until he was comfortable with the pitch of the engine and the RPMs it generated, and wouldn't touch the throttle again for twenty-four hours. We chugged along at six knots per hour, about the pace of a fast jog. When Hajo needed a break, I took the helm. Nothing to it, I learned. Just aim the bowsprit at some convenient landmark—a shore light or radio tower would do—or, better yet, apply the oldest of navigational tactics and steer toward a star or constellation low in the sky. Of course, stars march steadily westward, so every now and then I had to check the compass to make sure we weren't sliding toward the edge of the earth.

Behind us lay Lake Michigan, ahead Lake Huron. The Straits of Mackinac are wide enough, at five miles, to allow the two lakes to merge freely. All the other lakes are connected by rivers that fall over the slopes and ledges of the continent to the Atlantic. Superior, standing at about 600 feet above sea level, tumbles 23 feet to Huron. Huron slips 8 feet down to Erie, and Erie takes a big jump, 326 feet, into Lake Ontario, most of it over Niagara Falls. Only Lakes Michigan and Huron share a level.

Under sail or motoring, Hajo could rarely keep still. It's why he stays slim. When I relieved him at the helm, he moved over to sit on the rail and relaxed for a moment, scanning the water ahead. He jumped to his feet and went below deck and returned with a handful of peanuts. He sat again on the rail, shelled a few peanuts and popped them in his mouth, offered me some, jumped to his feet again and walked the length of the boat, leaning over to make sure no lines dragged in the water. He returned to the stern, tinkered with the flashlight taped to the compass, shelled a couple more peanuts, and went below to scan the gauges. He climbed to the poop deck, sat again on the railing, looked around, ate another peanut, and leaped up to check the compass heading. Then he went below and gathered the navigation charts in his arms and brought them on deck and spread them open on the cabin roof, weighting their corners with binoculars and flashlights.

Around us the water remained looking-glass calm, a mirror image of the moon and stars. I imagined throwing a stone out there and watching it surprise an exclamation mark from the surface—a tiny spout rising, concentric rings fracturing the moonlight.

Both times I'd boated through the Straits the wind was down—unusual luck, considering how often the wind blows there. A friend of mine, Richard Vander Veen, who is co-founder of an energy company called Bay Windpower, told me that the Straits are among the windiest places in the Great Lakes. His company is raising five wind turbines on the south shore, near Mackinaw City. The turbines will stand on 220-foot towers and their 100-foot blades will generate 1.65 megawatts of electricity each (1 megawatt is enough to run about three hundred average households). One turbine will supply the electrical needs of Mackinaw City's water-treatment plant, and the other four will power the town.

I didn't have the quote at hand or I would have read to Hajo what Father Jacques Marquette wrote about the wind in the Straits. Marquette is the best remembered of the early Jesuit missionaries, and by all accounts was a good human being and a reliable observer. He left France for the New World in 1666, at the age of twenty-nine, and until his death from illness in 1675, had an enormous impact on the Great Lakes region. After studying Indian languages for two years at the mission in Trois-Rivières, he was sent to Sault Ste. Marie, where he established the first mission. A year later, in 1669, he transferred to La Pointe mission on Lake Superior's Chequamegon Bay. In 1670, fears of an attack by the Sioux drove the local Ottawas and Hurons east, and the mission was closed. Marquette returned to the Straits of Mackinac and established a mission at St. Ignace. Three years later, he and fur trader Louis Jolliet, along with five voyageurs, would set out to find a route to the great "Michissipi" River that was said to lie west of the Fox River country. By then many Europeans were convinced that the mighty river they had been hearing about flowed to the Sea of California and would open a trade route across North America. Marquette and his party canoed across northern Lake Michigan to Green Bay, ascended the Fox River, portaged to the Wisconsin River, and rode it downstream to the Mississippi. They paddled about 650 miles down the big river before fear of Spaniards forced them to stop near the mouth of the Arkansas River, still 700 miles short of the sea. But they had seen enough to conclude that the river flowed into the Gulf of Mexico, not the Pacific. On their return trip to Mackinac, they discovered the shortcut to Lake Michigan via the Illinois River and the Chicago Portage.

Before and after his exploration of the Mississippi, Marquette stayed on Mackinac Island. In his journal he wrote about the effects of wind on the native fishermen he often watched netting whitefish and lake trout from the surrounding waters:

> The winds occasion no small embarrassment to the fishermen; for this is the central point between the great lakes which surround it, and which seem incessantly tossing ball at each other. No sooner has the wind ceased blowing from Lake Michigan than Lake Huron hurls back the gale it has received, and Lake Superior in its turn sends forth its blast from another quarter. Thus the game is played from one to the other; as these lakes are of vast extent, the winds cannot be otherwise than boisterous, especially in the autumn.

Most days the Straits are streaked with whitecaps, the wind constant and strong. Often I've driven over the bridge when every car and truck leaned in a sailboat cant. Once or twice drivers have been killed when they lost control of their vehicles in high winds and jumped the railing, plunging to the lake below. I didn't want Hajo to get the wrong idea.

"It isn't usually like this," I said.

But he knew that. Every sailor is alert to the hazards of straits and narrows. In salt water, a strait like Mackinac would create a tidal race churning with current and with whirlpools large enough to swallow small boats. Here, where tide is negligible, wind and cold are the dangers. The wind funnels between the headlands of the Straits, building seas that can drive a boat aground on the rocky shoals around every island and along both shores. Even a minor accident can be fatal in water that rarely warms above sixty degrees. Ice clogs the Straits from January through March most years, often closing it entirely. Shipping continues only as long as the Soo Locks remain open at the outlet of Lake Superior in Sault Ste. Marie. Most years the locks are closed from about mid-January through March. Until then, a channel is kept clear through the Straits and in the St. Mary's River by the *Mackinaw*, which at 290 feet long is the largest icebreaker stationed on the Great Lakes. It is moored at the Coast Guard station in Cheboygan, fifteen miles east of Mackinac City. During mild winters, the cutter can keep the channels open until spring; but most years the cold wins, and by deep winter, ice has closed the upper lakes.

When the *Malabar* passed through the Straits of Mackinac, she was passing through some of the richest history in the Great Lakes. The Straits were the thriving center of the Old Northwest, an immense territory extending from the Ohio River in the south to Hudson Bay in the north to the Rockies in the West. In all the early French and English references to the Old Northwest, two places stood out: Michilimackinac and Sault Ste. Marie. Both were settled by the French in the 1660s and 1670s, and soon became the westernmost headquarters of the soul-and-fur industries.

The Straits, where Huron and Michigan merge, are only fifty miles from Lake Superior, creating the tightest conjunction in the Great Lakes system. This close grouping of lakes gave the area strategic and commercial value for thousands of years before the arrival of Europeans. The Straits were beyond the reach of the Iroquois, who controlled much of the land east of Lake Huron and south of Lake Erie, and thus became a traditional meeting place for councils of tribes from as far away as the Great Plains. They continued to gather there long after the Europeans arrived with their Bibles, steel hatchets, iron kettles, flintlock rifles, and rum. The Indians were usually indifferent to the Bibles but became quickly dependent upon the rest. It was only a matter of time before they were willing to trade land for commodities.

The heart of the Straits is the island of Michilimackinac. The name translates variously, but usually to something like "Big Turtle," for its tortoiselike shape. The Indians considered it holy and buried their dead in its limestone caverns, but the French and English valued it because its promontories made it more easily defended than the low mainland.

Among the first white men to see the island was Jean Nicolet, who passed through the Straits in 1634 on his way to Green Bay. Claude Dablon, a French Jesuit priest who built the first mission on Mackinac Island in 1670, called it "The key and the door for all the people of the south." For hundreds of years it would be the commercial and military capital of the Great Lakes.

Settlements formed at three places in the Straits. The earliest was on a point protruding from the north shore, where Marquette established the mission he called St. Ignatius. The French built a fort there about 1678, then abandoned it three years later. The town of St. Ignace now is the northern terminus of the bridge and a jumping-off point for ferries to Mackinac Island.

By 1715, the French had built a new, larger fort on the south shore of the Straits, at present-day Mackinaw City. The original stockade was located on the beach (where a reconstructed fort and museum now stand), and was exposed to wind and shifting sand, with the forest at its back. In 1761, after the defeat of the French in the French and Indian War (begun in 1754), the British took possession of the fort, training its cannons on the Straits and controlling the fur trade. But in 1763, the Ottawa chief Pontiac engineered an uprising of tribes against the British and attacked fourteen of their forts, including Michilimackinac. "Pontiac's War" was brilliantly orchestrated. Indian men gathered outside the fort to play "baggatiway," a traditional racquette-and-ball game similar to modern lacrosse, and one of the players knocked the ball over the wall into the fort. As the men ran into the stockade in pursuit of the ball, women handed them tomahawks they had concealed under their robes. The warriors slaughtered all but a few of the soldiers in the garrison. Not until the defeat of Pontiac a year later did the British regain control of the fort.

In 1780, the British built a new and more easily defended fort on Mackinac Island and kept a garrison there even after the American Revolution, finally handing it to the Americans in 1796. They recaptured the island during the War of 1812, but relinquished it for good in 1815. It became a national park in 1875—the nation's second, after Yellowstone—and a Michigan state park in 1895.

That night on the water, I was struck by how little the Straits have changed. The bridge has altered it, of course, and the clustered lights of Mackinac Island, Mackinaw City, and St. Ignace are prominent. But most of the shores, seen from far enough off, can appear as wild and uninhabited today as they must have to Nicolet and Marquette. Even Mackinac Island, despite its fudge and T-shirt shops, has in some ways changed little in two centuries. Some of that is deliberate—no motor vehicles are allowed, for instance. But even the population has remained about the same. In 1818, the island had five hundred year-round residents and in summer swarmed with thousands of fur traders and Indians who gathered to rest and refresh themselves and trade with local merchants for trinkets and drink. Two centuries later, the permanent population is still about five hundred. And in summer, thousands of tourists swarm the island to rest and refresh themselves and trade with local merchants for trinkets and drink.

On our starboard, larger than Mackinac Island and dense with forest, was Bois Blanc Island, or "BOB-lo." Every March for several centuries much of the population of Mackinac Island rode dogsleds the couple miles across the ice to Bois Blanc and camped there to make maple syrup. For two or three weeks they tapped the island's sugar maples, collected the sap, and boiled it communally in large brass kettles. It was a highlight of the year, a time to celebrate the end of winter and to indulge a craving for sweets. Some years the ice in the Straits rotted and broke up before syruping was finished. When that happened, everyone on Bois Blanc had to wait a week or two for the ice to clear before they could be rescued in boats.

One day in 1787 a man named Charles Patterson, a partner in a fur-brokering company, set out from Mackinac Island in a birchbark canoe loaded with eight voyageurs and a young Indian woman described only as his "slave girl." Also aboard was Patterson's large white dog. They paddled west from the Straits, hugging the north shore, bound for Green Bay. Forty miles into the journey, the canoe apparently swamped. Bodies were found a few days later, washed up on shore at a place known to this day as Point Patterson. Only the dog survived.

Nobody knows how many people have died in these waters. Their canoes were engulfed by waves or split open on rocks. Their sailboats ran aground during storms and they climbed the rigging to get away from the waves and froze to death there. They dozed on watch and their vessels collided with other vessels and went down. They slept in their bunks as boilers exploded or fires broke out or bulkheads staved in. Booms swung unexpectedly and knocked them overboard. They got drunk and walked off decks. They stood on gunwales to urinate and lost their balance. They set off walking across the ice and disappeared forever. They shouted, "I'll get it!" and dove over-board to retrieve a lady's feathered hat and never came up.

On September 20, 1856, a 110-foot brig named the *Sandusky* was bound for Buffalo from Chicago carrying a cargo of grain and a crew of seven. That night she was caught in a gale in the Straits and broke up in high waves. Some of the crew climbed the masts and were clinging to a spar as another vessel passed. A side-wheeler, *Queen City*, set out from Mackinaw City to try to reach the sailors, but in the high waves and gale-force winds they could not get close enough to save them. The cold claimed them, and they fell one after another into the water and disappeared.

Late on the night of June 26, 1878, the schooner *St. Andrew* collided with the schooner *Peshtigo* five miles west of Cheboygan. In those days before radio and radar, collisions were among the most common causes of wrecks. Many thousands of vessels were registered on the lakes in the last half of the nineteenth century, and ten thousand would pass through the Straits every year. Now and then their paths intersected. Often the damage was slight, and no lives were lost. But that night in 1878, both vessels quickly sank. The crew of the *St. Andrew* made it into their yawlboat and were saved, but two of the *Peshtigo*'s crew drowned.

On the morning of April 20, 1909, the 213-foot steamer *Eber Ward*, downbound from Chicago with a cargo of 55,000 bushels of corn, came around Gray's Reef, turned east into the Straits, and plowed into an ice floe. Five crew members went down with her.

In dense fog on May 7, 1965, the *Cedarville*, a 588-foot lake freighter carrying a cargo of 14,411 tons of limestone, collided with a Norwegian freighter about three miles from the Mackinac Bridge. With his ship taking water through a large hole in her hull, the captain steered her for shore, hoping to run aground at Mackinaw City. But two miles offshore, in 105 feet of water, the *Cedarville* suddenly rolled on her side and sank, taking ten of her crew of thirty-five with her. The others were saved by lifeboats from a nearby vessel.

Such carnage was hard to imagine that night on the *Malabar*. I looked across the glassy water at the lights of Mackinac Island and thought how glad we would have been the summer before for a dose of storm—a very small dose, just enough to spur *Gauntlet* ahead to the finish line. But on the *Malabar* we were grateful for the calm.

Chapter 6

LAKE SUPERIOR

If we'd had the freedom and the time, we could have steered the *Malabar*
north from the Straits and entered a channel between the mainland Upper
Peninsula and Drummond Island called Detour Passage, where the St.
Mary's River empties into Lake Huron. There we would have found a
sprawling, slow river a mile or more wide. We could have wound among
darkly wooded islands forty-five miles up the river to the city of Sault Ste.
Marie and the rapids of the St. Mary's. Sault Ste. Marie is two cities: one
in Michigan and a larger one across the river in Ontario. The "sault" or
rapids is where Lake Superior slips over a ledge of broken bedrock into the
river and roars a half mile downhill before the water slows and deepens.
For many centuries before the first Europeans arrived and for another century or two after, "the Soo" was a traditional gathering place for Indians of
many tribes, who would come to parley and trade and especially to fish in
the rapids for whitefish. Early travelers had to portage around the rapids,
but since 1855 a complex system of locks and canals has bypassed them. If
we had entered the locks, we would have been lifted thirty feet in elevation
and would have emerged at the edge of Superior, the biggest lake of all.

But we could not go north. We didn't have that luxury, and it stirred
worries in Hajo that were never far from his mind. He was fifty, time was
flying, every day counted, there were many places he wanted to go before
he died. Superior was just one more on a list so lengthy it saddened him.

But it also made him determined. Already he was planning a return visit with his own sailboat, the 61-foot *Argonaut*, a Sparkman and Stephens modeled after Irving Johnson's *Yankee*. He wanted to sail *Argonaut* from Duluth, at the western extreme of Superior, through the Soo Locks and down the St. Mary's, through the Straits and down Lake Michigan to Chicago. From there he would take the canals that lead south to the Ohio River, then to the Mississippi and down to the Gulf of Mexico. But first he had a job to do.

"This is a delivery," he kept reminding us—and himself. Steve Pagels wanted the *Malabar* in Maine as quickly and cheaply as possible so he could put her to work to start earning back his investment. We knew we shouldn't blame him for being practical-minded, but we did. For Hajo, it was frustrating to be so near the largest of the lakes and unable to see it.

So that night, as we guided our boat around the top of the horseshoe, I told him a little about it. Or tried to. Superior is too big for easy explanation. In surface area it is the largest freshwater lake in the world; only Baikal in Russia has more depth and greater volume. Superior is larger, deeper, cleaner, colder, less developed, and less traveled than the other Great Lakes. It is also more deadly. At 350 miles long, 160 miles wide, and with a surface area of 31,700 square miles, it is capable of swallowing oceangoing vessels with shocking ease.

The French called it *le lac supérieur*, not because of its size or because they thought it had higher value than the other lakes, but because it was the northern or uppermost of them. Before them, the Ojibwa called it *Kitchi Gami* or "Great Lake." To the explorers seeking passage to the Pacific and to the fur traders and missionaries who followed, Superior was a road they could follow deep into the continent. From its shores they ascended tributaries that led even deeper, all the way to the Hudson Bay country to the north and the Great Plains and Rockies to the west. Much of its north shore and portions of its other shores have changed little since the days of the voyageurs. And though not free of environmental problems, Superior has remained the most pristine of the lakes.

I told Hajo about living in Marquette in the late 1970s, while attending Northern Michigan University. From the hilltop house my wife and I shared with our friends Dick and Trisha Armstrong (who live there still), we could look out over Marquette Harbor and a hundred and some square miles of Lake Superior. The year-round surface temperature of the lake averages just forty-three degrees Fahrenheit, and in winter it always hovers near freezing, but constant wind and waves usually prevent ice from forming except in bays. But the winter of 1977–78 was extraordinarily cold. The wind blew bitter and hard from Canada and coated the highway with sheet ice. Snow piled to the eaves of houses and entombed cars in their driveways. We skated down the street to the supermarket, and partied through the fiercest blizzard of our lives. For weeks the temperature barely rose above zero. One morning it was thirty-five below zero at the shore, fifty below a few miles inland. For the only time on record the lake froze all the way across, forming a sheet of ice nearly as big as the state of Maine.

In milder seasons we would sometimes walk the shore between the immense loading docks on the waterfront. Pelletized iron ore from the mines near Ishpeming and Negaunee was carried by train to the loading docks and dumped a carload at a time into the vast echoing holds of lake carriers up to a thousand feet long, each capable of carrying 25,000 to 60,000 tons of ore. When the holds were full, the boats set off for Gary or Detroit or Lorain or Cleveland, their cargo destined for the steel mills of the industrial Midwest and East.

Superior has always reminded me of charismatic people, the ones who dominate a room by their presence. Like all the Great Lakes, it creates its own weather, but it creates more of it and across a larger area, dramatically affecting a significant portion of North America and influencing the attitudes and lifestyles of people in three states and a Canadian province.

I told Hajo that the lake's water is so cold that only in rare summers can people swim comfortably in it. It is so clear that you can see rocks on the bottom more than thirty feet down.

Jonathan Carver, a New Englander who set off in 1766, at age fifty-seven, to explore the Upper Lakes, wrote about Superior in *Three Years Travels Through the Interior Parts of North America*: "When the weather was calm and the sun shone bright, I could sit in my canoe and plainly see huge piles of stones beneath six fathoms of water. My canoe seemed to be suspended in air. As I looked through this limpid medium, my head swam and my eyes could no longer behold the dazzling scene."

Six fathoms is thirty-six feet. Most of the lake is as clear today as it was in 1766.

I told Hajo about Isle Royale, the national park near the top of Superior, where people thousands of years ago came to chip raw copper from nuggets the size of automobiles. I told him about the "Witching Tree," a gnarled white cedar growing at the tip of Minnesota's Hat Point at the west end of the lake—a tree the Indians consider holy and that is so old it was described in the journals of French explorers two and a half centuries ago. I told him about the ancient artists who painted shoreline rocks with ocher representations of moose, bear, and mythical lake monsters. I tried to make the lake seem real to this man who had sailed all his life on oceans. I wanted him to know Superior was a major body of water and not to be trifled with.

I told him that while I was a student in Marquette, I went off alone one weekend in search of new places. Near Munising I hiked to a remote section of the shore and camped on a slab of rock so close to the water that the spray of waves nearly reached me. I spent two days and nights there, tending a driftwood campfire that barely pushed back the cold. Mostly I watched waves batter the rocks with hammer blows. They went at it with great patience. For a hundred centuries they had been carving stone into shapes suggestive of water itself.

Trains of waves and random breakers rose and fell as far as I could see. I imagined being out there in a storm—climbing the face of every wave, cresting it, then plunging into the trough, wave after wave like that, each impact ripping bolts from wood, tearing metal apart, splitting seams in the hull. Finally, horribly, the boat would plunge into a trough and not come up. Maybe there would be a last moment of hope—the desperate promise that because you were in your bunk or clinging to a mast, you were safe. And then the crush of water, the shock of cold forcing all the air from your lungs, the boat going down beneath you. Your terror would probably be brief, at least. The cold would drag you into sleep so deep breathing became irrelevant.

At some point that weekend I concluded that the lake's indifference to my welfare was more terrible than any malevolence I could imagine. It had greater implications. I, who had assumed such significance for myself, had none. How could it be, I wondered, that this bizarre accumulation of atoms and molecules that I called my self, and the inexplicable spark that animated it, mattered so little? Where would the spark go when I was dead? What proof would there be that I had ever lived?

It was not the first time I'd considered such matters, and certainly not the last, but it made a powerful impression. I shared the story with Hajo because I assumed he must have had similar realizations many times while sailing the oceans. Who better than a sailor for pondering the mysteries? When better than during a night sea journey?

"I watched those waves for two solid days," I said, "and for the first time realized that the dark secret of life, the one that parents are careful to keep from their children, is that the universe doesn't give a damn. It seems obvious now, I know, but if you grew up like I did, thinking of nature as a friendly grandmother who serves you blueberries in milk when you're hungry and tucks you into a feather bed when you're sleepy, it came as a shock. Grandma turns out to be a heartless bitch. She doesn't even blink as you slip shrieking beneath the waves."

Hajo stood huddled beside the helm, his face hidden by the hood of his parka. He said nothing for a few moments. I thought he must be wondering why I hadn't mentioned God. Where was God in this story—in any story?

"When you sail in the Caribbean," he said, "be sure to always take the berth on the low side of the boat."

"Why's that?"

"Because of the cockroaches." He pronounced it "cock-a-roaches."

"What about them?"

"The cock-a-roaches are really smart. They hate salt water, so they always climb to the high side of the boat, and they'll eat your underwear, the bastards."

I laughed. "I didn't know that," I said.

"Keep it in mind the next time you're in the Caribbean."

. . .

I could have told Hajo other stories. About Superior's north shore, for instance. I'd explored it several times during road trips to hike the beaches and fish from shore, but had never been there in a boat. In May a few years ago I decided to see it from a canoe.

Spring is the season of calm on Superior. Winter's cold remains impounded, damping the wind. Often the lake is covered with fog—thick white rafts of it, as big as New England states. You don't want to be on the water then. Stay on shore, tend a campfire, read a book, whittle. After the fog lifts and while the wind stays down, the lake makes a good highway for canoes.

But not a busy highway. More small boats used Superior two centuries ago than they do now. In 1800, more than a thousand voyageurs a year paddled from Montreal to the head of Superior and back. Though most of the land surrounding the Great Lakes has been altered beyond recognition, Superior's north shore has changed little. For mile after mile there are no human signatures of any kind. Contained within Lake Superior Provincial Park, Pukaskwa National Park, the bulky protrusion of the Pukaskwa Peninsula, and the Nipigon country are roadless forests and the least visited and most rugged shoreline in the Great Lakes. Moose, black bear, and wolf live there, and remnant populations of woodland caribou. It is still *la marge sauvage*—"the wild shore." You can see glimpses of time's broad scale there. A paddler in 1800 saw the same remote shore lined with the same billion-year-old rocks. He saw the same crystalline waves breaking on the same cobbled beaches, saw the same spruce forest rising away to an interior wilderness so vast it seemed infinite.

I was there as a guest of David Wells, owner of Naturally Superior Adventures in Wawa, Ontario. Wells is a fit and relaxed forty-two-year-old who knows the north shore intimately. He invited me to meet him at his home and business, located on a stunning rocky point at the mouth of the Michipicoten River near the town of Wawa, and join him and ten other paddlers for a three-day paddle in a 36-foot replica of a voyageur's canoe. We would camp on beaches very likely used by the voyageurs themselves.

Though we would be in a fiberglass canoe, not birchbark, and would employ a few modern conveniences, including a two-way radio in case of emergency, it was to be a trip steeped in tradition.

I asked David about the wisdom of paddling an open canoe on a lake that can devour seven-hundred-foot ships. His answer was predictable but reassuring: "If you treat it with respect—and pay attention to the weather—it's safe."

That night, I slept in a tepee. It had been erected on the sandy shore of the Michipicoten, close enough to Lake Superior so I could hear the surf all night. The river here was a slow widening before the lake. Fur traders used the Michipicoten to get to the Hudson's Bay Company posts far to the north near James and Hudson bays. The river is slow much of its length—an advantage when paddling upstream. At its top is a height-of-land portage of two hundred yards to a lake at the beginning of the Hudson Bay drainage. From there you can portage to the Missinaibi River, paddle over the edge of the Canadian Shield (carrying around the spectacular waterfalls there), and ride the current north to Moose Factory, on the shore of James Bay. I lay in my sleeping bag in the center of the tepee, the tall ridge poles towering above like a forest falling inward, and fell asleep to the sound of breaking waves.

Early in the morning, I joined a group of a dozen people gathered on the shore of the river. Most of them were members of the Voyageur Trail Association, a volunteer group working to complete the Northern Ontario Hiking Trail, which will eventually trace the entire north shore of Superior. The canoe trip was a raffle prize for them. The plan now was to explore the shore from the Michipicoten to the Dog River and back, forty miles round-trip, a short day's jaunt for the voyageurs. It would take us three days.

David's employee, Chris Mortimer, stood to address us. He was thirty-nine years old, lean and excitable, with the manner of a man who has successfully reinvented himself. He wore an embroidered voyageur shirt, a red sash around his waist, deerskin leggings, and moccasins. I would learn later he grew up on Lake Ontario's Bay of Quinte, an Ontario resort center, and was a regular attendee of workshops by Tom Brown, the wilderness survival guru from New Jersey. He made us understand that survival was serious business. He, for instance, ate only wild meat—no domestic beef or pork—and in a pinch could survive on a gruel of lichen scraped off rocks. He reminded us that the Earth is our Mother. That Lake Superior will kill us if we're stupid or disrespectful. That in the bush, time is irrelevant, which is why he doesn't own a watch. That he and David were the leaders and we must always defer to their experience and wisdom and not go running off on our own acting like crazy people.

He urged us to circle him, link hands, and join in a pagan prayer of thanks. Thanks, Lake, for giving us water to drink and a road to paddle. Thanks, Sun, for warming us. Thanks, Moon, for lighting our way at night. Thanks, Birds, for singing so prettily in the morning. Thanks, Fish, for letting us eat you. Thanks, Mother Earth, for all the usual reasons.

Then it was time to placate the Manitous. I remembered that the voyageurs, coming out of the French River (probably through the outlet now known as Voyageur Channel) and turning north up the coast of Georgian Bay, encountered open water as they rounded Grondine Point. There they would toss an offering of tobacco to the lake and say, *Souffle, souffle, la Vieille*—"Blow gently, blow gently, Old Woman." Chris, a fervent traditionalist, produced a bag of Bugle loose-leaf tobacco and sprinkled some on the water. A few of the others had tobacco, but most pawed around in their backpacks and came up with trail mix, raisins, and individual serving boxes of Corn Flakes and Lucky Charms. I appeased the gods with my breakfast granola bar—but I was hungry and could spare only a pinch.

The canoe rested in the sand. It was enormous, thirty-six feet long by six feet wide, big enough to haul a team of football players and all their equipment. We lined up six on each side and half-dragged, half-hauled it to the water. When it was floating, we formed a brigade and passed wanigans, duffels, and backpacks to Chris and David, who arranged them between the seats and low in the boat to keep it trim. One after another we stepped over the stern and scrambled forward, settling two abreast on hardwood seats, paddles in hand. Most of the paddlers were middle-aged, about half men and half women, and most had at least some experience in canoes. Chris took position in the bow as lead paddler, or *avant*. David, the *gouvernail*, stepped into the stern and manned the long steering paddle.

We pushed off. A few strokes carried us into the river's slight current. Chris leaned far over the gunwale and made plunging drawstrokes to turn the bow. The boat swung heavily downstream and we picked up momentum. We passed over the turbulent shoal at the mouth of the Michipicoten and entered the big lake. The wind was light, from the east, blowing a gentle chop. Low swells passed under us. David steered the boat west, lining it up parallel to shore.

"*Preparé!*" called David and we raised our paddles. Then "*En avant!*" and, in unison, we stroked the water. At first the canoe was reluctant to move. We could feel every pound it carried—more than three thousand of them by our estimate. But a few hard strokes broke the grip of the water, and a few more brought us to cruising speed. At forty strokes per minute we made four miles per hour, the canoe cutting easily through the water. Synchronization was not difficult. We needed only to match the rhythmic rise and fall of Chris's paddle—regular as a metronome, easy as heartbeat—to reach accord, twelve paddles flashing at once. The lift, the forward reach, the dip, the draw, the lift again—it was nearly effortless. If one or two of us at a time paused to stretch or get a drink of water, the rest of us noticed no diminishment. It began to seem strangely as if the canoe were paddling itself.

Canoes were the vessels of choice on Lake Superior long before the first Europeans arrived in the New World. The Ojibwa have a legend that their ancestors once lived along the Atlantic Coast and that during a time of terrible sickness a comet appeared in the western sky. Soothsayers claimed the comet was a sign directing them to a better land, one where they would find food growing on the water. The people gathered their belongings into their canoes and set out in search of this wondrous place. They paddled up the St. Lawrence River and discovered the Great Lakes. They followed the coasts of what we now call Lakes Ontario, Erie, and Huron to the St. Mary's River and portaged the rapids into Lake Superior. They followed the shore westward until, near the present site of La Pointe, Wisconsin, they found marshes filled with wild rice. Food on the water. They had reached their new home.

Many centuries later, the first French explorers arrived in the New World. In 1534, while crossing in his ship from Anticosti Island to the Gaspé Peninsula, Jacques Cartier made notes in his journal about the semicircular bay to the north, not realizing that this "bay" was actually the mouth of a great river. Near the shore of the Gaspé, Cartier met a party of three hundred Iroquois and made note of their birchbark canoes. He traded trinkets for furs and retreated across the Atlantic.

The following year he returned with three ships and named a bay along the north shore after St. Laurens, the patron saint of that day, August 10. He then sailed into the river that would eventually bear the same name, establishing a settlement at what would later become Quebec City, and proceeding as far upstream as the future site of Montreal, where rapids blocked his way. From that toehold on the St. Lawrence, the French would explore westward. By 1580, their ships were appearing every summer at the junction of the St. Lawrence and Saguenay rivers to trade with Montagnais Indians who had traveled downstream by canoe from the interior. The French were quick to realize that the birchbark canoes of the natives were the best craft for exploring the new land.

In 1603, when Champlain reached the rapids at the future site of Montreal, he was told by Indian interpreters that three vast lakes lay to the west. The most distant of them he assumed to be the Pacific Ocean. In 1608 he established a settlement and fort at Quebec, where twenty of twenty-eight settlers died during their first winter. Champlain explored the region south of the St. Lawrence in 1609, ascending the Richelieu River into what is now New York State and discovering the lake (sometimes claimed as the sixth Great Lake) that still bears his name. In 1613, he canoed north as far up the Ottawa River as Allumette Lake, following a route that had been pioneered by a young Frenchman, Nicolas de Vignau, two years earlier. Champlain went much farther up the Ottawa in 1615, to the Mattawa River, then to Lake Nipissing and down the French River to Lake Huron— a seven hundred–mile journey that for the next two centuries would be the favored route to the Great Lakes.

Champlain had probably been preceded into the interior by only one other European, his protégé Etienne Brulé, whom he had sent the previous year to learn the native languages and find a route to the western ocean. Brulé was the first of a new breed of French explorer, the *coureurs de bois* (literally, "rangers of the woods"). He and others like him lived among the Indians, learning their customs and dialects, traveling in their bark canoes, practicing their woods skills. They would spearhead much of the European exploration—and exploitation—of what would become known as the Northwest Territory. The Great Lakes stood at the heart of it.

The first white men to travel Lake Superior's length and record their experiences were two renegade fur traders named Radisson and Groseilliers (known to generations of Canadian schoolchildren as "Radishes and Gooseberries"). They canoed deep into the region in 1659 and 1660, departing Sault Ste. Marie on the south shore of Superior, spending the winter trading for furs at Chequamegon Bay, at the southwestern end of the lake, and returning along the north shore. From the Soo they took the French and Ottawa rivers to Quebec, returning to the capital at the head of a fleet of sixty canoes carrying a fortune in furs—$300,000 worth, which represented most of the revenue New France would earn that year. After two years in the wilderness, they were convinced that most of the fur wealth was to be found north and west of Superior, making the difficult and dangerous French/Ottawa route obsolete. They proposed that it would be more efficient to extract that wealth via the rivers flowing north to James and Hudson bays, where sailing vessels could be used. But the governor of New France was not impressed. Radisson and Groseilliers had failed to obtain a license before heading into the wilderness and had angered influential officials by refusing to allow Jesuits and government agents to accompany them on their journey. Instead of being welcomed as heroes, as they expected, they were arrested. Groseilliers was jailed, both men were fined, and their pelts were confiscated. Embittered, they appealed to England, where they found support from businessmen eager to compete with the French.

In 1668, Groseilliers sailed the forty-ton *Nonsuch* into Hudson Bay and wintered at the Rupert River on the southeast shore of James Bay. He returned to England with the hold stuffed with furs. Trade in the north country would never be the same. British investors lined up, a royal charter was granted, and in 1670 the Hudson's Bay Company was formed. Soon, trading posts—"forts" and "factories"—were established on the Moose and Albany rivers in James Bay. Fourteen years later, the English stronghold was expanded with the construction of York Fort at the mouth of the Hayes and Nelson rivers on the west shore of Hudson Bay. The Hudson's Bay Company had tapped into the richest beaver country in the world.

The French realized too late that they had blundered in letting the English get established in the north. They tried to reach James and Hudson bays by an overland route, but the terrain was too difficult. One party of voyageurs and a Jesuit completed the journey in 1671 and 1672, but only after making two hundred portages and negotiating four hundred sets of rapids. After the French and English went to war in Europe in the 1680s, hostilities broke out in Hudson Bay as well. A war of skirmishes and ambushes and seizures of furs and property lasted sporadically for nearly thirty years. The rivalry between the French trading posts on the Great Lakes and the English posts in the Hudson Bay country would last 150 years, until fashions for beaver hats—and the beavers themselves—were exhausted.

But while the trade lasted, it flourished. From the late seventeenth century to the middle of the nineteenth century, the voyageurs made constant forays to the Great Lakes region to withdraw its wealth of animal pelts. Every May, brigades of canoes left Montreal. Most were birchbark cargo canoes thirty-six to forty feet long. Called Montreal canoes or *canots du maître,* they were manned by crews of twelve or sixteen voyageurs and often carried several Jesuits and other non-paddling passengers. Each canoe was loaded with three tons of trade goods such as axes, knives, guns, traps, brass kettles, wool blankets, and rum or brandy. The brigade paddled up the Ottawa River to the Mattawa River, made a short portage over a headland to a small tributary of Lake Nipissing, crossed the lake, and descended the French River to Georgian Bay. From there the voyageurs coasted north to Mackinac Island. Or they continued north up the St. Mary's River, portaged the rapids at Sault Ste. Marie, then hugged the north shore of Lake Superior for more than five hundred miles to the western end of the lake, ending their trip at Fort William, in Thunder Bay, or at Grand Portage, at what is now the boundary between Ontario and Minnesota. The trip took eight weeks.

At Fort William and Grand Portage, the voyageurs from Montreal were met by voyageurs who had traveled by rivers down from the northwest in smaller, 25-foot *canots du nord*—"canoes of the north"—loaded with beaver pelts. The meeting of the two groups was celebrated with a week or two of revelry. As many as two thousand voyageurs would party and rest during these annual rendezvous. Then they switched cargos, the *hommes du nord*, or "men of the north," returning inland to spend the winter trading with native trappers for more pelts, and the *mangeurs du lard*, or "Pork Eaters," retracing their route to Montreal, arriving with their loads of furs before the waters froze in late fall.

The industry earned fortunes for the merchants in Montreal and Europe. For the voyageurs themselves, it was a life of adventure and danger—and brutal hours, backbreaking labor, and little pay. Perfect, if you had an abundance of spirit and not many options and were too young to know better. And what adventurous young man of eighteen or twenty would know better?

So we were modern voyageurs, paddling steadily along the north shore of Superior in a fiberglass canoe painted to look like birchbark. Occasionally David called for a rest and we would drift close to walls of rock sculpted by the waves into stairways, patios, and palisades. David drew our attention to diabase dikes, where molten rock once poured through cracks in the bedrock. He pointed to pillowy black Precambrian rocks as big as buses. Embedded within them were smaller stones that a billion years ago were thrown into the sky during volcanic eruptions and landed in soft magma. The magma hardened and the missiles remained embedded, like chips in cookies. When the glaciers crept down from the north, they sheared the tops off the ancient rocks, slicing them smooth. The stones within them were sheared off also, leaving the black surface looking like it had been inlaid with gems and polished. We were silent then, aware of the great age of the rock and the power of water and ice. The lake rose and fell gently beneath us. It was like resting on the back of a sleeping animal.

Few places in America's middle are as geologically vivid. When the Swiss geologist Louis Agassiz toured here in a canoe in 1848, he was fascinated by deep scores in the bedrock and by the many large rocks he found that somehow had been transported to the lake's shore from hundreds of miles north. Previously, most of the earth's geologic features had been attributed to floods, especially the biblical Deluge. Based partly upon his observations along Superior, Agassiz proposed that the force that altered much of the surface of North America and Europe was actually ice. In a long-ago Ice Age, he concluded, massive glaciers scored rock and churned the land like "God's great plow."

Our day was lovely: temperature in the high seventies, a following breeze, the sky as clear and blue as Superior itself. A few miles west of the Michipicoten, all signs of habitation ceased. No houses along the shore, no radio towers on the hills, no evidence that anyone had been there before us.

We stopped for lunch on a sandy beach at the mouth of the Doré River, a twenty-foot waterfall in view upstream. While the others built a fire and unloaded cooking gear, I got out my fly rod and fished for steelhead. Third cast a seven-pound hen whacked my bright streamer, then fought lethargically, like her thoughts were elsewhere. I landed her, admired the scarlet stripe and silver sides, and released her back into the river.

We paddled all afternoon. The day grew warmer, even in the band of cool air that hung above the water. We dipped our cups in the lake and drank from it.

The north shore of Superior is among the few places in the Great Lakes where you can safely drink the water. With less than one percent of the world's surface water fit to drink, a reservoir the size of Superior is priceless. And not only in the aesthetic sense. Over the years a number of plans, schemes, and literal pipe dreams have been suggested as ways to profit from Great Lakes water. In 1981, a pipeline company in Montana sought a permit to run large-diameter pipes from Lake Superior to the coal country of eastern Montana. The company wanted to pump water from the lake to Montana, where it would be mixed with crushed coal to form a slurry. The slurry would then be piped back to loading docks in Duluth, for shipment to world markets.

The Montana coal slurry proposal was rejected, but it raised caution flags.

A year later, the governors of five of the Great Lakes states and the premier of Ontario met in a conference on Mackinac Island to discuss the issues of water sale and diversion. The delegates were unanimous in their determination to "keep it clear and keep it here," as a later governor of Wisconsin would phrase it. But the journalist William Ashworth noted in his book *The Late Great Lakes* (1986) that the meeting was conducted with the assumption that the water in the lakes would probably someday be for sale to the highest bidder. In answer to the suggestion that water could become the OPEC oil of the future, former Ontario premier William Davis asked, "Have you ever considered what the Great Lakes would be worth at $25 a barrel?"

More recent events have raised the same question. In 1998, a Canadian company headquartered in Sault Ste. Marie was quietly granted a permit to export up to 156 million gallons of Lake Superior water every year for five years. The company intended to pump water from the lake into tanker ships for transport to overseas markets. The news created an instant uproar. Governors and representatives of the Great Lakes states drafted letters requesting the permit be revoked, and a bipartisan group of congressional delegates introduced a resolution in the U.S. House of Representatives asking the president and the Senate to block all sales of Great Lakes water. The Ontario minister of the environment promptly canceled the permit and issued a press release that read, in part, "Ontario is determined to make certain that water from the Great Lakes never appears on anyone's commodity trading board."

But clearly there are those who would put it there.

Late in the afternoon we stopped to make camp on a sandy beach at the base of a tombolo, a peninsula formed when an island close to shore links to the mainland with a sandbar. The resulting landform is shaped something like a mushroom on a stem. The stem makes a fine place to camp because one side or the other is always sheltered from the wind.

I gathered driftwood for a fire. The beach was narrow, made of fine, dark sand. At its back were high ledges of rock, and above the ledges rose impenetrable walls of conifers. Both up and down the shore the ledges fell to the water's edge. This was the only sandy place to be seen, and I had no doubt that it had been used as a campsite for thousands of years. When Chris kneeled in the sand and lit a campfire by striking a chunk of flint with a piece of steel, I experienced a moment of historical vertigo.

Dave nudged me and pointed at the sun. It was surrounded by a halo, a portent of rain in both ancient folklore and modern meteorology. Sunlight high in the atmosphere was being prismed by ice crystals. To the west, cirrocumulus clouds were visible just above the horizon. I thought of the east wind that had followed us all day. An east wind often precedes the anticlockwise gyre of a storm system.

"How long, do you think?" Dave asked.

"Two days," I said. But I was kidding myself.

"More like twelve hours."

A rule of thumb followed by sailors on the Great Lakes is that it takes three days for a storm to build and three days for it to blow itself out. Pierre-François-Xavier de Charlevoix, the Jesuit priest, canoed the north shore of Superior in 1720 on a journey that carried him ultimately to the mouth of the Mississippi. On the subject of Superior's storms, he wrote:

When a storm is about to rise you are advertised of it, say they, two days before; at first you perceive a gentle murmuring on the surface of the water which lasts the whole day, without increasing in any sensible manner; the day after the lake is covered with pretty large waves, but without breaking all that day, so that you may proceed without fear, and even make good way if the wind is favourable; but on the third day when you are least thinking of it the lake becomes all on fire; the ocean in its greatest rage is not more tossed, in which case you must take care to be near shelter to save yourself. . . .

That night, Dave, Chris, and I decided to sleep voyageur style, spreading a large tarp on the sand, stretching it around the hull of the overturned canoe, and staking it overhead to form a roof. It seemed like a good shelter until late in the night, when a wind came up and the canvas started bucking, filling the air with sand. I grabbed my sleeping bag and ran to my little dome tent, which I'd left set up as a precaution. The rain came. It fell hard—probing, jabbing, searching for a way inside and often finding it. A sudden buffeting of wind pounded the walls of the tent and made it huff and puff like an indignant squat ogre.

When I stepped from my tent in the morning, the bay was flurrying with racehorses and the wind was cold. I climbed the bank at the foot of the tombolo and faced windward, into stinging rain laced with sand. As far as I could see were whitecaps jumping from the tops of six-foot waves. When Superior throws a tantrum, it's hard to accommodate the scene. You shield your face from the wind and try to look through your fingers at all those miles of surging waves and whitecaps. You see clouds scudding in retreat, see gulls beating their wings into the wind, hear the battlefield sounds of surf—and your perceptions shut down. The scene is too big. It bludgeons your senses. After a few minutes you need to turn your back on it and concentrate on something smaller.

We were stormbound. *Dégradé*, the voyageurs called it.

All day we sat together beneath the tarpaulin in our down jackets and wool hats. A harsh, spitting rain turned to downpour, then slackened, then downpoured again. The canoe on its side formed a dark dry cubby in the back of the shelter. Up front it was wetter, the rain driven under the tarp by the wind, but we built a smokey campfire there and put on a kettle to boil, and that's where most of us gathered. We sat on the wet tarp facing the lake. Tattered gray clouds drudged along above the whitecaps. Once an updraft grabbed the tarp and uprooted the poles and ropes from the sand. The wet canvas collapsed upon us. After that we kept it elevated with our paddles.

As people have always done in similar situations, we told stories. Jean-Paul Dubreuif and his sister Dorice had spent most of their lives in Quebec, but were living now in Sault Ste. Marie. When they talked to each other, it was often quietly in French; their voices carried the intimacy of a secret language from childhood. They repeated folk tales and sang French-Canadian ballads they had learned as children.

Most of the others were born and raised near Lake Superior. Verna Scott remembered the isolation of the lake's second largest island, Michipicoten, not many miles west of where we now sat. In her childhood, her uncle lived alone there as the lighthouse keeper, and Verna and her parents spent part of every summer on the island with him. "It was awesome and wonderful," she said. Once her uncle found the body of a man washed up on the beach. He buried him in the woods. They never learned who he was.

Dan and Loretta Sweezey lived on Batchawana Bay and owned a 26-foot sailboat in which they often cruised the north shore. Dan was retired now, but for years he worked as a wheelsman on lake carriers, making regular runs from Duluth to Montreal. For a couple years he steered a tug towing logs along the north shore—massive two-mile rafts of spruce and tamarack cut from the interior for the paper mills in Sault Ste. Marie. In the lake the loose logs were encircled in a gigantic noose of boom logs linked together with chains. Sometimes logs broke free or were washed from the boom, and freelancers in tugs would return and gather them one by one and sell them back to the paper mills. If the wind was strong from the east or northeast, Dan's tug would churn in powerful futility, losing headway, and was sometimes pushed back all the way across the lake until the boom broke to pieces against the Michigan shore.

Bruce Hanna told the legend of the snow wasset. This terror of Lake Superior was as big as a schooner and looked something like a cross between a giant snake and a whale, with heavy scales on its belly and a layer of blubber beneath its skin. It was said to be capable of swallowing a man in a single gulp. In bad weather the wasset would follow a ship for days, waiting for a chance to sink it and devour its crew. The Ojibwa spoke of a similar mythological lake serpent they called "Mish-ie-was-set-ta-gan-ie-gan-nis-sa," a name folklorist Ivan H. Walton translates as "a very big monster and you're a goner if you meet one."

At thirty years old, Bruce was the youngest of our group and by far the largest. He was a looming hulk of grinning good nature, a former country boy from British Columbia hungry to know the world. He had recently finished ten years of university study in environmental sciences. He could have finished sooner, but he was too restless to sit long in classrooms and kept taking off on journeys to Europe, Asia, Australia, and Central America. He had always aspired to be "a traveler, not a tourist," so whenever he found himself in a place that caught his interest, he stayed a while. Sometimes a long while. For now he was living in a boardinghouse in Sault Ste. Marie and earning starvation wages, a virtual stipend, from the Ontario Department of Natural Resources, Fisheries Division. He assisted field researchers, netting brook trout from inland lakes, for instance, to check their growth rate and health. The hours were long, the work physical and often uncomfortable, but it was an opportunity to fatten his résumé and make him more marketable in the competitive field of Canadian natural resources. He wanted to do original research of his own someday, perhaps on Lake Superior. But first he needed to travel more. South America was pulling at him.

After lunch, Bruce and I decided to hike a mile and a half down the shore to a small river we had passed the day before and fish for steelhead. We set off carrying backpacks containing sandwiches, fishing gear, a map, and matches. Chris insisted we take along his emergency whistle, a precaution I thought unnecessary.

"You'd never hear it over the waves," I said.

"I might if I was looking for you."

The beach petered out quickly, replaced by rock ledges that dropped abruptly into the water. At first we could walk on the ledges, but they became tilted and the rain made them slippery. The entire shore was scalloped with small, deeply vaulted coves, each with rock walls twenty or forty feet high rising sheer from the water. Waves, funneling into the coves, tried to climb their sides and launched in spray to the sky. We clambered higher and began cutting across the base of every peninsula, scrambling over rocks, using small trees for handholds, clawing our way into the interior.

In the woods, day gave way instantly to dusk. Everything was dripping and gleaming from the rain, and we were soon soaked, even through our raingear. The forest along the Superior shore has a jungle thickness to it. Abundant fog, rain, and snowfall support a richness of trees and shrubs, a panoply of ferns and mosses. The woods at the eastern and western ends of the lake are mixed with deciduous trees and conifers. But creeping down from the north to the midsection of the shore, from our stretch near Wawa and nearly to Thunder Bay, is boreal forest—black spruce, balsam fir, tamarack, and hemlock—that has grown here since it succeeded treeless tundra and dwarf birch 9,500 years ago. Stretching from northern Maine to northern Minnesota and far north into Canada, this is a forest region ecologists sometimes call the "spruce-moose biome." Boreal vegetation thrives on the scanty layer of soil covering the Canadian Shield, and moose thrive on the vegetation.

Chris had told us that he and some of his neighbors in Wawa were in the habit of rising before dawn and patrolling Highway 17, looking for moose that had been struck by cars and trucks the night before. A few times a year he finds a freshly killed bull, cow, or calf and butchers it on the spot, filling the back of his compact car with hindquarters and rib steak. Moose are famous for strolling down the highway at night, where the going is easier than in the shin-tangles. I've seen them many times—a glitter of eyes higher off the ground than seems natural, and suddenly a massive black shape is towering broadside in the road looking at you. A moose stands so tall that its legs are often taken out from underneath it by a sedan, causing it to fall through the windshield. In this part of Canada a moose can weigh over a thousand pounds. It's wise to drive with vigilance.

Bruce and I were surprised to find moose signs in the thickest woods along the shore. How could such large animals, with such broad antlers, make their way through this mess? The trees entwined their branches from crown to ground, tangling in an understory dense with blueberry bushes, Labrador tea, laurel, and leatherleaf. Sphagnum moss grew knee-deep in clean, bright mats as soft as feather beds, disguising rocks and concealing crevasses. Tussocks of grass, thickets of young evergreens, fallen trees smothered with moss, sudden narrow chasms like rips in the hide of the earth with ferns clinging to their sides and streams rushing fifteen feet below—it was tough walking. The only way to get through was to shove, dodge, duck, and wend, pushing ahead blindly and letting the forest slam shut behind us. In five minutes we could have been lost forever. Without sun or compass there would be no hope of maintaining orientation. Without the sounds of the lake to guide us—and a hundred feet into the forest the boom of surf was absorbed to silence—we wouldn't know if we were walking toward Lake Superior or Hudson Bay. But we had brought compasses.

And we were lucky that the season was too early for mosquitoes and that the rain kept the black flies down. In another week or two the blood-pests would arrive in their billions.

After three hours, we had gone less than a mile. If we continued we could not get back to camp before dark. The prospect of spending the night in that wet forest was so unappealing that we abandoned the fishing trip and returned more or less the way we had come. It took another three hours to go back. The next day we would travel the same distance in the canoe in about fifteen minutes. That was against a headwind, and in no particular hurry. We realized that woodland Indians invented the birchbark canoe because they *needed* it.

That evening under the big tarp, we diced and peeled and grated and stirred until upon each of our plates appeared that staple of backcountry cuisine, the burrito. We drank a pretty good merlot from a box. For dessert Chris stirred together apples, pears, raisins, and pecans, simmered them with sugar and cinnamon, and rolled the steaming medley in tortillas. Then more wood on the fire to heat a kettle of lakewater for tea, and more stories and jokes and songs.

Finally, eagerly—for we were wet and chilled—we retired in ones and twos to our tents. We lay in the twilight and listened to the wind roaring overhead and the rain spattering in gusts and the earth-trembling boom of waves hitting the oldest rocks on the planet. We could hear the crawl of centuries out there.

We didn't make it to the Dog River and Denison Falls. The lake set the agenda, as it always does. The storm broke on the third day and we returned to Michipicoten the way we had come, staying close to the spectacular, raw, humbling shore, out of the wind as much as possible, the rocks and trees still dark with rain and appearing altogether different from the ones we had seen under blue sky.

At noon we stopped for lunch at a shallow bay paved with so many smooth, round cobblestones that it looked like they had been dumped there by the bargeload. It was difficult to walk on them, impossible to find a comfortable place to sit, until we dragged driftwood logs into a circle. Chris cleared enough stones to build a fire and put a kettle of soup on. Dave picked his way down the beach, then veered into the trees. A few minutes later he reappeared and shouted for us to come.

A few paces inside the woods was a weathered picket fence enclosing a rectangular plot measuring eight feet by three. The fence must have been carried there and erected at great inconvenience. Though we found no head-stone, we decided only a human grave would be worth so much trouble. The fence was old, with just a few traces of white paint still visible on it, and a black spruce twelve inches in diameter grew from the center of the plot. We wondered if this was the resting place of a drowned sailor or an Ojibwa fisherman or a crewman fallen from an ore boat. We thought of the body Verna's uncle had found on Michipicoten Island. I recalled a story told by a couple I visited once who were living in the lighthouse on Lake Michigan's Beaver Island. The lighthouse had long before been decommissioned, but they stayed there as caretakers of the property. While out walking in the woods, they had discovered a wooden cross hidden in bushes. Curious, they contacted the previous lighthouse keeper and learned that he had found the body of a man on shore in the 1950s and buried it there. He hadn't known what else to do with it.

Maybe this grave had a similar history. Maybe someone had found a body washed up on the beach and buried it here in the woods, above the reach of ice and waves.

We stood quietly for a while. It seemed to me that digging such a grave would be an act of obligation, duty, or simple decency. Any of us would do it. But building the fence, that was different. That was compassion.

Chapter 7

LAKE SUPERIOR

The Gales of November ◆ The Sinking of the *Edmund Fitzgerald*
◆ A Survivor's Story

November is the deadliest month. Ask any sailor. It's when the lakes still embrace some of summer's heat, but the air above has turned to winter. A meteorologist for the National Weather Service once calculated that on average the greatest difference between the temperature of the lakes and the temperature of the air above them occurs on November 10. That differential causes the remaining warmth in the lakes to be sucked into the air, releasing energy and creating wind. With so much energy available over such large bodies of water, even minor storms are intensified. Nobody knows how many ships have sunk in the Great Lakes—estimates range from four thousand to ten thousand—and nobody knows how many of them have gone down in November, but the number is high. The danger to ships is magnified because November is the end of the shipping season, and companies trying to wring a little more profit from the year will sometimes take greater risks than usual. Many ships have been wrecked on what was scheduled to be their final run of the season.

Several November storms are legendary, and most of them occurred on or near that critical tenth day of the month. The great storm of 1913 began on November 7 and continued through the 11th. Often referred to as "the Big Blow," it was the result of a freakish collision of weather fronts. One low-pressure cell of cold air roared down from Canada, while another swept east from the Rockies, absorbing heat as it passed over the Great

Plains. The two fronts met over Lake Superior, absorbed additional energy from a third system rushing north from the Caribbean, and produced a massive storm that engulfed all five Great Lakes. Hurricane-force winds and waves as high as thirty feet sank twelve ships (eleven of them with all hands), damaged twenty-five others, and drowned between 250 and 300 mariners.

The Armistice Day storm of November 11, 1940, was probably as fierce as the Big Blow of 1913, but it was concentrated mostly over Lake Michigan and did not take as many lives. November 10 that year was sunny and warm, with a light easterly wind, and the forecast for the next day, Armistice Day, promised to be just as pleasant. But far to the west, an intense low-pressure cell packing powerful winds was sweeping inland from the coast of Washington. It struck the Tacoma Narrows Bridge with such force that the bridge oscillated violently and finally collapsed (an incident caught famously on film), then continued eastward, passing over the northern Rockies and across Montana and the Dakotas. At the same time, another low-pressure system charged down the flank of the southern Rockies, absorbed warm, moist air from the Gulf of Mexico, and set off northeast across the southern plains. When the two systems collided, they formed what is arguably the most powerful storm to strike the interior of North America in the twentieth century.

The U.S. Weather Bureau announced a storm warning for the Great Lakes at six-thirty on the morning of November 11, while the lakes were still eerily calm. Many ships took the forecast seriously and stayed in port. Others had already left. Some ignored the warning and went on with their business.

Over Iowa, high winds, hail, and cold temperatures flattened buildings and killed livestock. Barometers in Duluth, Minnesota, Houghton, Michigan, and La Crosse, Wisconsin, were recording their lowest readings in history. By midday the storm hit Chicago, toppling chimneys and billboards, stripping roofs off buildings, lowering the lake level along the waterfront by five feet. On the east shore of the lake, seas grew at a rate that veteran sailors had never seen equaled. By early afternoon, winds up to a hundred miles an hour had produced waves estimated at thirty-five feet. Ships caught on the open lake had little chance of reaching safe harbor.

Lost with all hands near Ludington, Michigan, was the *William B. Davock*, a 420-foot bulk carrier bound for Chicago with a crew of thirty-two. Lost also was a 320-foot Canadian vessel, *Anna C. Minch*, and her crew of twenty-four. The 253-foot *Novadoc*, with nineteen crewmen and a cargo of powdered coke, was driven onto a sandbar near Ludington and broken in half by the waves; seventeen of her crew were rescued, but two were washed overboard and drowned. In all, fifty-eight mariners died on Lake Michigan that day. Also lost were fifty duck hunters who went to marshes along the lake early that morning and froze to death in their blinds or drowned when their small boats were overwhelmed by waves.

On November 18, 1958, a 640-foot lake carrier named the *Carl D. Bradley* met a storm in northern Lake Michigan while northbound from Chicago, where she had discharged a load of limestone. Not far from Beaver Island she was overtaken by winds from the southwest at sixty-five miles per hour and by waves twenty to thirty feet high. Because the *Bradley* was running empty, she was high in the water and therefore took a greater beating than normal. At five-fifteen in the afternoon the captain radioed a routine message to the home port in Rogers City to say all was well and they would arrive on schedule. Fifteen minutes later, the ship began to break in half. In another fifteen minutes she sank to the bottom in 350 feet of water, taking thirty-three men with her. Two survivors were rescued fourteen hours later by the Coast Guard, after clinging to a life raft in those twenty-foot seas.

Other deadly storms occurred on November 28, 1966, when the 660-foot ore carrier *Daniel J. Morrell* broke into two sections and sank in heavy seas in Lake Huron, with the loss of all but one of her crew of thirty-two; and on November 10, 1975, when the *Edmund Fitzgerald* was lost with all hands on Lake Superior.

Even without the Gordon Lightfoot ballad it inspired, the *Fitzgerald* would be the most famous Great Lakes shipwreck. At 729 feet long, she was the largest vessel on the lakes when she was launched in 1958, and for the next seventeen years was probably the most widely recognized. The *"Big Fitz."* was the flagship of a new generation of modern lake carriers. Nobody imagined that she would ever sink.

On November 9, 1975, the *Fitzgerald* shipped out from Superior, Wisconsin—Duluth's sister city in the southwestern corner of Lake Superior—with a cargo of 26,116 tons of taconite ore pellets, bound for Detroit. Her crew of twenty-nine was considered one of the best on the lakes, and her captain, Ernest McSorley, was a veteran of forty-four years of shipping.

Gale warnings were announced a few hours after the *Fitzgerald* departed port. A weather system had begun generating over the Oklahoma panhandle the day before and was heading north. Eventually it would grow into a monster swirling counterclockwise across much of the eastern third of the United States, from Minnesota to the eastern seaboard. But early indications were for a much more modest storm. Winds on Lake Superior were expected to reach thirty-eight knots, waves five to ten feet. Lake sailors sometimes call a storm of that magnitude a "fringe gale." They're common in November, and are not usually severe enough to keep large ships in port.

Around two in the morning of the 10th the forecast was upgraded from a gale warning to a storm warning. By that afternoon, the wind was blowing in excess of eighty miles an hour with gusts to one hundred miles an hour and waves had reached thirty-five feet.

On the *Fitzgerald*, several things went wrong in quick succession. Not far east of Michipicoten Island, her long-range radar failed. She still had short-range radar, but it could not provide a clear indication of the vessel's location. Captain McSorley knew approximately where he was, but he needed to know exactly. He had hoped to stay close to Caribou Island to get some protection from the waves, but he had to stay clear of Six Fathom Shoal, an area of rocky shallows just northeast of the island. Meanwhile, wind knocked out power at the Whitefish Point Lighthouse, blacking its light and ending transmissions from its directional beacon. A gasoline generator that served as an automatic backup failed to start. Without radar and with no beacon assistance from shore, the *Fitzgerald* was virtually blind.

The storm continued to build. As the *Fitzgerald* passed precariously near

Six Fathom Shoal, the Coast Guard sent an emergency broadcast directing all ships on Lake Superior to find safe anchorage. In Sault Ste. Marie winds were being clocked up to ninety-six miles per hour and waves were washing over the gates of the Soo Locks, forcing them to be closed. Wind was gusting to eighty-five miles per hour on the Mackinac Bridge; it too was closed.

About ten miles behind the *Fitzgerald* was the *Arthur M. Anderson*, a bulk carrier that had stayed in regular radio contact and provided navigation assistance when the *Fitzgerald*'s radar failed. Through breaks in the storm the two vessels could sometimes see each other's lights. The *Anderson*'s captain watched the *Fitzgerald* pass dangerously close to Six Fathom Shoal. Soon afterward, he received a radio message from McSorley reporting that the *Fitzgerald* was listing to starboard. McSorley had turned on his bilge pumps, he said, and was reducing speed to allow the other ship to catch up.

Sometime around 7:00 P.M., while the *Fitzgerald* was about fifteen miles from the entrance to Whitefish Bay, the first mate of the *Anderson*, aware of the listing of the other vessel, asked McSorley over the radio: "How are you making out with your problem?" McSorley answered: "We are holding our own." Twenty minutes later a blinding snow squall passed between the two ships. When the snow cleared, the *Fitzgerald* was gone.

More than a quarter century later, the cause of the *Edmund Fitzgerald*'s sinking continues to be debated. Five expeditions to the wreck site with manned and unmanned submersibles (and at least one free dive with scuba equipment) have failed to explain the accident. The official Coast Guard board of inquiry report concluded that "massive flooding in the cargo hold" was probably to blame. Ineffective hatch closures were cited as the most likely reason for flooding.

Many people disagreed. The Lake Carriers' Association initiated its own investigation and argued that the steel hatch covers used on the *Fitzgerald* were of a type proven effective during more than thirty years of use. Based on radar readings of the vessel's heading, damage observed, and especially Captain McSorley's several radio reports that his vessel "had a bad list" to the starboard and that he had ordered ballast pumps put into operation, the association concluded that the ship hit bottom at Six Fathom Shoal near Caribou Island. Damage to the hull, they argued, would have flooded the ballast tanks on one side of the vessel, accounting for the list.

Another theory is that the vessel broke apart on the surface, the result of stress fractures caused by flexing and twisting in the heavy seas, the same forces that broke the *Bradley* and the *Morrell* in half. Some people have also suggested that the final blow might have been impact from the "Three Sisters," a quasi-mythical sequence of anomalous waves that have often been reported on Superior. Most descriptions of the waves say they are unexpected and enormous, much larger than any others encountered. While a vessel wallows under the impact of the first two Sisters, a third arrives, the Big Sister, and delivers the *coup de grâce*.

Early one morning in November 1998 I woke in my bed at home and knew instantly that something was wrong. It was the wind. In the seven years my family and I had lived in our house on Old Mission Peninsula, near the Lake Michigan shore, I had never heard a wind like it. It made a steady howling roar, like a herd of animals in a panic, and at intervals seemed to double in volume. The windows banged in their frames. The entire house seemed to lurch and shudder. Beyond the wind I could hear waves pounding the shore of East Bay.

A television meteorologist reported that gusts at the lighthouse at the tip of Old Mission were measured at sixty-seven miles per hour, and increasing. If they reached seventy-five miles per hour, that would be considered hurricane strength.

It was November 10, the peak of the stormy season on the Great Lakes, and the twenty-third anniversary of the sinking of the *Edmund Fitzgerald*. I grabbed a sleeping bag and a camera and set off for Lake Superior.

On the Mackinac Bridge my truck shook under the impact of gusts and leaned so hard to the left that I had to keep the steering wheel cranked a half turn right to compensate. When I'd called bridge authorities from home, they were less than reassuring: "The bridge is *not* closed at this time," a woman told me, "but I can't guarantee that it will be open when you get here." On Mackinac Island, gusts had reached eighty miles per hour. Later they would be clocked at ninety-five. I joined a short line of traffic and followed a police cruiser across the bridge at fifteen miles an hour.

Below us, the Straits were wracked with storm. Whitecaps in long, frothy striations charged from east to west. It was bad down there. A lake carrier had taken refuge in the lee of Mackinac Island. Five others were clustered behind Bois Blanc Island. All had dropped anchor and swung their stems into the wind. They were massive—seven-hundred-footers, I think—and looked absurdly out of place huddled behind the islands.

Electricity had failed across most of the Upper Peninsula and the roads were deserted. I drove cautiously north, branches vaulting like tumbleweeds across the pavement in front of me. Whole trees had been uprooted or snapped off at the stump and had fallen across the road. Maintenance crews bundled in foul-weather gear and hardhats were chain-sawing some of them away, but new ones kept falling. I drove slowly, in four-wheel drive, edging into the ditch to get around the fallen trees. Most of the towns along the route seemed abandoned, their restaurants and service stations empty. Only the corner taverns were open, each with a few cars and trucks in front. Through the windows I could see candles burning on the bartops.

North of the village of Paradise the road follows closely along the shore of Whitefish Bay. The wind blew strong from the east here, unimpeded across the water from Ontario. Waves came in parallel lines of froth, break-ing far offshore and rushing the beach, their tops stripped off by the wind. They thundered onto the shore, climbed as high as they could, and re-treated, taking huge chunks of the beach with them.

At the end of the road, at Whitefish Point, is the Great Lakes Shipwreck Museum, a cluster of buildings around the base of the towering Whitefish Point Light Station. Across the parking lot is the visitors' center for the Whitefish Point Bird Observatory. The Point is a migration funnel for birds crossing Lake Superior. In the spring and fall birders from all over the world flock here to train their binoculars on the thousands of raptors, waterfowl, and songbirds streaming overhead.

Both the observatory and museum were closed, so I walked past them, leaning into the wind, and followed a sandy trail to the shore. The beach at the Point is broad and sandy, cobbled near the water with smooth, multicolored stones thrown up by centuries of waves. Large logs of driftwood, bleached silver and tumbled smooth, lie half-buried in the sand. The wind was cold, the air filled with sand and spray.

Whitefish Bay is big—nearly thirty miles wide and thirty miles long. Across its width, in Ontario, rose ranks of hazy mountains. To the east, beyond sight, was Sault Ste. Marie. To the west, Superior opened wide and seemed to go forever.

Whitecaps covered all the lake. They were bright beneath the dark sky. Most of the waves closer to shore appeared to be eight to ten feet high, but larger ones towered above them. A chaotic, angry mix of breakers and rollers attacked the beach. The waves had no discernible pattern until they neared shore. Then they organized themselves into rows that climbed higher, then curled, paused a moment, and broke with impact so powerful the ground shuddered. There was regularity to them—a row of breakers every three seconds, like clockwork, raising a tumult. It took a few minutes to realize that some of the noise was caused by thunder. A bolt of lightning speared the lake half a dozen miles away.

Away to the west, not far from where the *Edmund Fitzgerald* went down, a giant bulk carrier rode low in the water, bound for Sault Ste. Marie. A diagonal smudge of rain overtook it, erasing it from view.

I returned to my truck and drove a half mile to a small harbor on the east shore of Whitefish Point. A dozen commercial fishing boats were tied to docks, riding out the storm. The wind here was so strong that it made my truck rock in the parking lot. I sat inside, wipers and defroster fan turned as high as they would go, and watched the lake going crazy.

A steel-and-concrete breakwall circled the harbor. It was taking the full brunt of the storm. Each wave struck the face of the concrete and launched a plume of spray two or three stories high. The wind caught the spray, carried it in horizontal sheets a hundred yards inland, and slammed it against my windshield. It was like sitting in a car wash. I tried to take photos through the glass, but it was hopeless.

Then I saw something astonishing. A man was standing on top of the breakwall. I watched in disbelief. He was dressed in dark clothing and carrying something bulky in his arms, equipment of some kind that he struggled to hold as he bent against the wind. A wave exploded against the wall and the man turned his back to it and huddled down. The plume launched high, dwarfing him, then fell, obscuring him from view. When the spray cleared, he took a few steps along the wall and bent over his equipment again.

For fifteen or twenty minutes I watched, expecting any moment to see the man disappear into the lake. Finally he turned and hurried the length of the breakwall to shore. As he came closer, I realized he was wearing a diver's wet suit and carrying what appeared to be a large video or motion-picture camera. I stepped out of my truck into blasts of wind and caught up to him at the back of his pickup truck in the parking lot.

His face was red from the wind and cold, his eyes bright with excitement. Shouting to be heard above the wind, he explained that he worked for the Shipwreck Museum and was gathering footage for a documentary they were making about Great Lakes storms.

"Are you here for the reunion?" he shouted.

"What reunion?"

"The *Edmund Fitzgerald* reunion. Every year the family members come to the museum for a memorial. You're welcome to come. I'll take you there."

People arrived at the museum bearing platters of cookies. Every time the door of the main building opened, a rush of wind and rain came in. The power was still out, so dozens of candles had been placed in plastic lids and distributed around the room. In their flickering light, brass artifacts and glass display cases glittered everywhere. On display were models of lost ships, a carved eagle from a steamer sunk in 1892, the ship's bell from the schooner *Niagara*. The walls were lined with enlarged photographs of divers exploring Great Lakes wrecks. High on a pedestal, an enormous Fresnel lighthouse lens dominated the room. A staff member climbed a ladder and placed a kerosene lantern behind the glass. It cast a wavery glow across the room. The beveled edges of the lens scattered fragments of light on the ceiling.

Most of the hundred or so people who gathered in the museum that evening were relatives of the twenty-nine crew members lost on the *Fitz-*

gerald. There were wives and a few parents, many children and grandchildren. Most of them knew one another well. They'd been meeting on this date every year for twenty-three years. They shook hands and hugged, talked in low voices, laughed quietly.

Thomas Farnquist, the founder and executive director of the museum, walked to the front of the room and stood before a table displaying the ship's bell of the *Fitzgerald.* It had been salvaged during one of the early expeditions to the wreck. At the conclusion of the memorial, the granddaughter of a lost crew member would stand solemnly at the table and ring the bell twenty-nine times. But that would come later. Now, with a word from Farnquist, everyone took seats in folding chairs and went quiet.

The guest of honor that night was John Lufkins, the fifty-five-year-old administrator of the Bay Mills Indian Tribe. Most tribal members live on reservation land just down the lake from Whitefish Point, in or near the fishing village of Brimley. Lufkins walked to the front of the room and stood before us with his head bowed. He was gray-haired and dignified, wearing a suit and tie.

"I have a story to tell," he said in a quiet voice. "But it is hard for me to tell it. I haven't shared it with many people."

His story was not about the *Edmund Fitzgerald,* but about another drama that occurred November 10, 1975, the day the big ship sank. Lufkins was a thirty-two-year-old commercial fisherman then, living in Brimley and fishing with gill nets for whitefish and lake trout in Whitefish Bay. That morning, he and his brother-in-law, eighteen-year-old Pat Kinney, went to work like they did nearly every day.

They launched their boat—a sixteen-foot, open launch powered by an outboard motor—and went out on the lake to haul their nets. It was a typically cold November morning, the bay relatively calm. They motored far into Whitefish Bay, near tiny Tahquamenon Island, where they encountered some heavy seas, but nothing to worry about. They decided to begin hauling the nets that were farthest out, then work their way back to shore.

But the seas were growing and the temperature was dropping. By afternoon, the waves were so large that it became difficult to handle the heavy nets. Then it became impossible.

Other fishermen out that day remember seeing the entire lake suddenly boiling and churning. None of them had ever seen anything like it. Most storms start slowly and build. They give warning. But this one swept down upon them so quickly that they could not get off the lake. Roaring into the bay from the open lake came the largest and most violent waves Lufkins had seen in all his years on Superior.

The situation became critical. He and his brother-in-law could not possibly make it to shore. Their only hope was Tahquamenon Island.

They beached on the lee side of the island and pulled their boat above the reach of the waves. They knew they would have to spend the night there and they needed shelter. They got lucky. On the island were a pair of abandoned tarpaper shacks. The first was wrecked beyond use, but the roof and walls of the other were intact. Inside was a cot and an ancient oil stove. The stove's fuel tank held a couple inches of oil in the bottom. They managed to light it.

Lufkins kept a pair of sleeping bags in his boat for emergencies. He carried them into the shack, along with a thermos of hot coffee. He and Kinney settled in for a long wait.

Suddenly the door flew open and a fisherman they knew, Billy Cameron, stood in the doorway, soaked to the skin and nearly frozen to death. His boat had capsized, he said, and his partner, Andrew LeBlanc, was in the water. The three men ran to the shore and found LeBlanc struggling in the surf, clinging to an empty gasoline can. The three waded into the wash and pulled him to safety.

But the danger wasn't over. They had to take steps to prevent hypothermia—or "exposure," as it was still called in those days. Half-walking and half-carrying Billy and Andrew, they got the two men inside the shack, stripped them of their wet clothes, and wrapped them inside the sleeping bags.

It was late afternoon by then and the day was growing dark. Lufkins went outside to circle the island one more time, scanning the lake for others who might be in trouble. He saw nothing but enormous breaking waves to the horizon. He returned to the shack, planning to stay inside all night.

But as he opened the door, the wind ripped it from his hands and slammed it against the wall. He turned to grab it and when he did he caught a glimpse of color far out on the lake. There it was again: a flash of orange. A life jacket. Somebody was in the water out there.

Pat Kinney wanted to take the boat himself, but Lufkins wouldn't let him. Pat was young, he had his life ahead of him. Lufkins was the older and more experienced; it was his responsibility to go. He stripped to his long underwear and strapped on a life jacket. He and Pat went down to the beach and pushed the boat into the waves. He jumped aboard, started the engine, and turned the bow quickly into the next incoming wave.

Water rushed across his feet and the boat wallowed. He had forgotten to replace the drain plug in the bottom. He turned the boat to shore and ran it up on the beach and replaced the plug. Again they pushed the boat off and again Lufkins motored into the waves and powered through the line of breakers.

By now he could no longer see the man in the life jacket. He knew approximately where he had been, but in those seas, in such powerful wind, he had probably been blown far downwind by now. And it was getting impossible to see. A little daylight remained, but the sky was low and black, and the rain and spray were blinding.

Lufkins estimated the waves at fifteen to twenty feet. He would power the boat at full throttle up the steep slope of a wave and reach the crest, where he could get a glimpse of the surrounding chaos, but the wind would catch the bow of his boat and spin it around. He would motor into the trough, turn, and climb the wave again. Every time he reached the top, he was spun around and sent back down. In this way he made progress, but it was progress without control. He knew with heart-sinking certainty that the person he had seen in the water was going to die there.

But he kept trying. He figured that as long as there was a trace of daylight, he had a chance of finding him.

And then he powered over a wave and below him was a capsized boat with two men in life jackets clinging to the hull. They were family: his uncle Francis Parish and Francis's son, Christopher. They'd been in the water for two or three hours by then and were so exhausted they could barely raise their heads. Lufkins steered down the wave toward them. He figured he had one chance to save them, so he plowed his boat directly on top of the hull of the capsized vessel, between the two men. If they could grab his gunwales on opposite sides, maybe they could climb inside without capsizing him.

But the men were too weak to lift themselves. Lufkins crawled forward and grabbed Christopher by the back of his life jacket and somehow pulled him over the side and into the boat. He lay there, too exhausted to speak.

A wave struck them and the two boats separated. Lufkins scrambled back to the motor and accelerated it to turn his boat into the next wave. But the motor stalled. He cranked the starter cord. He cranked it again. It started and he turned the boat a moment before a wave would have struck him sideways and swamped him. He climbed to the crest, but the other boat and Francis were gone. He tried to remain in the same area, circling as best he could, accelerating to avoid breaking waves, shooting lengthwise down the troughs, turning quickly to bust over crests. He had no idea where he was.

Then, again, blind luck: the capsized boat was dead ahead, Francis still clinging to it. Again, Lufkins ran his boat unto the upturned hull. Again, he crawled forward and grabbed a fistful of life jacket. He hauled Francis over the side. The older man fell into the bottom of the boat. He seemed as limp as a rag doll.

Christopher raised his head from the bottom. His eyes were glazed with cold, most of the life and hope drained from him. He asked: "Are we going to live?"

"Damned right we are," Lufkins said. "I didn't risk my ass coming out here just so you could drown."

But in his heart he knew they were doomed. He had no idea where they were. It was almost night now and he could not see the island. No land was visible anywhere. All he could see in any direction were waves as big as hills, twenty feet high by now, one after another racing downwind so fast they overran themselves and broke apart in tumbling froth, only to climb higher than before and continue on, roaring and crashing.

He didn't know how many minutes they had before they ran out of luck. It would take just one wave to drown them. If the engine stalled again, they would die. If they ran out of gas, they would die. As wet and cold as they were, if they did not find shelter soon, they would die. He had no choice but to try to find that tiny island in the immense bay before the lake claimed them.

And there it was, dead ahead. They topped a wave and the island was right in front of them. Lufkins's brother-in-law and the others stood on shore, waving their arms. He shot across another wave and hit a trough as deep as a canyon and followed it to the breaker zone. A wave threw the boat into a rock and the motor sheared a pin and was useless. But they were close to shore now. Pat threw a line and Lufkins caught it. The men on shore pulled until the boat grounded on rocks, then lifted the two exhausted men out and carried them to the shack. They removed their wet clothes and wrapped them inside the sleeping bags. They gave them hot coffee. They would live.

Lufkins stepped from the boat and fell to the beach. His legs had gone to rubber. He knelt in the sand and gravel for a few minutes. "I felt," he said, "like the luckiest son of a bitch on the face of the earth."

It was a long night. They drank coffee and adjusted the stove to keep it burning. They tried to sleep, but couldn't. Once they heard a helicopter and ran outside into the darkness, but they had no way to signal it.

One of them had a small transistor radio. They tinkered with it, but the battery was almost dead and they couldn't pick up any stations. In the morning, they tried again and caught a newscast and learned that in the night, while they struggled to keep warm in their shack, an ore carrier twenty miles to the west was being obliterated in a snow squall—that while they drank coffee in stunned gratitude that they were alive, the *Edmund Fitzgerald* and her entire crew were plunging into the lake forever.

By now the waves had lessened, the worst of the storm past. Figuring it would be a long time before anyone found them, they replaced the sheared pin on the motor and pushed the boat into the water and climbed inside and headed for shore. By late morning they were home.

John Lufkins grew silent as he stood before the crowd of people in the museum. Nobody made a sound. The only movement was the flicker of candles. Finally Lufkins looked up and said, as if in apology, that he has never felt like a hero, which is maybe why it's been so hard for him to talk about what happened that day on Whitefish Bay. He said he felt as if some greater power took over and performed those acts for him and that he had no choice but to go along.

"It's just something I had to do," he said. "I don't know why. I don't know how I did it. But I think most people in that situation would have done the same thing. We lived, we got very lucky. But it breaks my heart that twenty-nine other guys weren't so lucky."

Superior is merciless. That night everyone in the museum knew it to the depths of their souls. John Lufkins will wonder all his life why it took mercy on him.

Chapter 8

LAKE HURON

In the morning we rounded the top of Lake Huron and steered the *Malabar* southward, wind at our backs at ten knots, a light chop on the lake. I straddled the bowsprit and listened to the prow turning water aside with a constant sound, like wheat pouring through a chute. This far offshore, the coast of Michigan showed as a low green fringe that made the lake seem even larger. To the north, east, and south only water and sky were visible. In the distance, a lake freighter rode low on the horizon, towing a smudge of smoke.

Lake Huron spread out around us. At 206 miles long and 183 wide, with a surface area of 23,000 square miles, it is the second largest of the Great Lakes and the fourth largest freshwater lake on earth. Like its sister lake, Michigan, it is deep (but not as deep as Superior), averaging 195 feet and with a maximum depth of 750 feet. The lake's most distinctive characteristics are its two enormous bays. Georgian Bay to the east is by far the largest bay in the Great Lakes, and Saginaw Bay on the west side is the third largest. It is the only Great Lake without a metropolis along its shores.

The radio claimed that the temperature of the air was sixty-five degrees on shore, but on the water we still wore insulated underwear and goose-down parkas. The northern lakes stay cold late into the spring and are cool all summer. One consequence of that is frequent fog. Another is optical phenomena. As I watched the horizon, it took on an icy shimmer, like mirages pooling on a highway. Spikes and stalagmites grew out of the shimmer and rose above the horizon line, revealing the wavering images of distant ships. A moment ago they were hidden beyond the curve of the earth; now they climbed into view, vivid but deformed, their smokestacks and pilothouses turned into turrets and crenellated walls. The air near the water was colder than the air above it and therefore denser. Light warped as it passed to our eyes had transformed ordinary ships into fantasy castles.

That afternoon, we raised sail for the first time. We had been making a little better than six knots all morning, a good average speed for the *Malabar*. With the sails up we still made a little over six knots, but the engine wasn't required to work as hard and we saved fuel.

The *Malabar* is two-masted, gaff-rigged, with topsails. During most of the nineteenth century, similar schooners were the workhorses of the lakes. They were preferred because of their relatively shallow draft and because gaff-rigged sails could be raised and lowered more quickly than square sails, and by a smaller crew. Sailors learned quickly how dangerous it is to run before the wind on the lakes, where the real hazard is land, not water. Run too far and you run aground. On the oceans, square sails once raised could stay up for days or weeks. Lake vessels did not have that luxury.

Raising the *Malabar*'s sails requires at least three people on deck—one at the helm, two sweating lines. "Sweating" is perfect jargon. The sails are heavy. You don't raise them easily. You must grasp a line as high as you can reach and throw your body away from it and down, your weight levering two or three feet of line through the pulleys near the top of the mast. You must do it quickly, with a cadence, to keep the sail climbing. At the same time, your crewmate draws the accumulating line around a belaying pin, to prevent it from slipping. The mainsail, heaviest of all, requires two people to raise; they stagger their grips on the line and sweat in unison, grunting mightily with every body heave. As the sail climbs, it catches wind and fills with a cargo of air.

Harold and I did much of the gruntwork. Tim helped if he could get out of the galley, but often he was cooking. Matt was a dynamo, throwing himself everywhere at once, sweating lines, belaying, bounding up the shrouds to unreef sails and disentangle lines, a knife in his teeth piratewise.

It was fascinating to watch, and Harold and I were eager to learn. But instruction of the green crew came only as an afterthought. The point was to get the job done, and there was not always time to explain procedures. Harold or I would be told to grab a line—"Not that one; *that* one!"—with no explanation as to the purpose it served. Matt needed our muscles, not our minds.

I fumbled with a knot while Matt hung from a line, waiting. He shouted at me to remove my gloves. Harold and I had already suffered damage to our hands and thought leather gloves were a splendid idea. Matt thought otherwise. "Grow your own, damn it," he growled, in the grand tradition of first mates versus deckhands. Catching the spirit of the tradition and indignant that Matt took for granted that my old carpenter's calluses had grown soft from deskwork (which they had), I threw my gloves hard to the deck and threw a muscular curse after them. Flagrant insolence. A century and a half ago it might have earned me a flogging. Now Matt fired off a dark look and growled with displeasure.

Matt's growl, I would learn, was harmless. There was about him a singlemindedness I came to admire. In conversation he was animated—but only while the subject was tall ships, navigation, wind, weather, or knots. Try to discuss anything outside his realm of interest and he listened impatiently for a few moments, his eyes skipping around, then at the first pause he directed the conversation back to a subject that mattered. He talked on and on, often about personal heroics in the service of tall ships, and often a bit longer than necessary. I suppose it was boorishness, but it grew from enthusiasm, and I found it easy to forgive.

Tim, however, could not forgive it. He and Matt had begun clashing even before we left Traverse City. On the boat they tried to avoid each other but of course it was impossible. "Hand me that belaying pin," Matt would say, and Tim would walk away. "Grab those dishes," Tim said, and Matt ignored him. Each was the master of a domain—Matt the deck and rigging, Tim the galley—and each informed the other at every chance. Once or twice early in the trip they squared off—Tim, pressing his height advantage, his pale thin face grinning with anger that looked almost like pleasure; Matt, shorter and stouter, his back and legs taut with muscles, face tanned dark and looking more bewildered than angry. They broke off quickly, for now.

Though Matt was only thirty years old, he was an accomplished sailor. He had grown up sailing competitively in Corpus Christi, Texas, and at seventeen years old won a full-ride sailing scholarship to the University of Hawaii. For two years he crewed on the university's 85-foot racing Maxi, *The Rainbow,* while earning an associates degree in physics. After school he stayed in Hawaii, getting a job on a high-angle rescue unit for the National Park Service. It involved being transported in helicopters and performing cliff and sea rescues of tourists fallen from cliffs, cavers lost in caves, climbers stuck on mountainsides, surfers dragged out to sea by rip tides. Too often he was sent to retrieve bodies, not survivors, and a disturbing number of them were children. He discovered a terrible truth: Fatalities occur in threes. A family would be standing near a fumarole, taking photographs of hell's steaming vent. One of the children would duck beneath a barrier and walk onto unstable ground. The ground would collapse, and the child would disappear into the earth. The father would go after the child. The mother would go after the father. Whoever remained alive would go for help. It would be up to the rescue unit to rappel into the pit and recover the bodies.

It was too heartbreaking for Matt, and he returned to sailing. He spent six months in London helping to rig a replica of Sir Francis Drake's *Golden Hind,* then sailed on her across the Atlantic on a four-year tour of the U.S. coasts. In 1989, while *Golden Hind* toured the east coast of Florida, Hurricane Hugo struck. As the storm approached, with winds as high as 138 miles per hour (they had reached 160 miles per hour in the Caribbean), the owners decided the ship's best chance was to sail out to sea. Matt and his crewmates took her into one of the fiercest hurricanes of the century and battled fifty-foot seas for five days before they returned safely to Cape Canaveral.

Later he worked on the *Santa Maria,* a replica built in 1992 to commemorate the half-millennial anniversary of Columbus's voyage, and on tall ships in Baltimore and Chesapeake Bay. He seemed to know every large sailing vessel in the world, its length and gross tonnage, its pedigree, its captain and first mate.

He liked to share his knowledge. Once after Matt and Hajo had conferred over the navigation charts, I asked, "What's the scuttlebutt?" He ignored the question but explained at length that the word "scuttlebutt" refers to the only place on the deck of old warships where sailors could talk without being overheard by officers. The "butt barrel" was where fresh drinking water was stored, and its opening was called the scuttle. It was the antecedent of the office water cooler. Sailors who gathered there inevitably passed gossip as they drank.

A "slacker," he told me, is a sailor shrewd enough to always grab the slack sheets and feed them out while someone else sweats the lines that raise the sails.

"Straight poop" is news delivered directly from the captain as he stands above the crew on the poop deck.

"Pull your own weight" means sweating your share of lines or pulling yourself up the shrouds.

The "bitter end" is the end of a line or "bit"—and when you've reached it, things often go badly.

"Three sheets to the wind" occurs on a ship only if something is very wrong with the rigging. On a square-rigger there can be only two sheets per sail, so the sail must have been rigged by someone drunk.

I soon recognized that Matt was an extraordinarily accomplished knotsman. Some people lose themselves in music, art, or literature; Matt loses himself in rope. He puts himself to sleep at night reading his deeply plumbed copy of *The Ashley Book of Knots*. Stored in the muscle memory of his fingers are knots to serve every possible purpose. Also many that have no practical use, but are decorative or simply entertaining to tie. Some were made obsolete by the invention of doorknobs, stepladders, Velcro, and brass zippers.

Often for relaxation Matt would sit on the deck of the *Malabar* with a length of hemp or nylon in his hands and practice sheepshanks, hitches, cringles, and bights. He had mastered the butt sling, the Chinese Crown knot, and the narrow Turk's head. Also the crown and diamond, the double crown and diamond, the double crown and Matthew Walker, and the double *single* Matthew Walker. Not to mention the double bastard weaver's knot. And the double twofold overhand bend—which, as far as I could tell, requires at least sixteen fingers and four thumbs to execute properly.

I found a short rope of my own and showed Matt the knot I learned so well in Boy Scouts thirty years ago. A square knot! Oops. "That's a whatnot," said Matt, but I knew it as a granny. Matt said it is also sometimes called a "false knot," or, by the unbearably pretentious, a *nœud de faux*. I had turned the end the wrong way around the bight. Fifty-fifty chance.

I tried a bowline, that most useful of sailor's knots, but wasn't even close. Matt, with utmost patience, showed me the moves—first the basic knot, then variations like the bowline hitch, bowline bend, bowline on the bight, and double bowline on the bight. He lost me with the variations, but I practiced the basic bowline for an hour. A month later I was getting it right about half the time.

Beyond sight across Lake Huron lay Manitoulin Island, the largest freshwater island in the world, where a deeply scalloped shoreline adds up to a thousand miles of coast. The island is speckled with a dozen towns; the largest, Little Current, has a population of fifteen hundred. Most of the residents live within the island's five First Nation reservations.

Past Manitoulin is Georgian Bay, an arm of Huron larger than Lake Ontario and almost as large as Lake Erie. Georgian Bay (and its long narrow top, called the North Channel) remains the wildest portion of the lakes south of Superior. It is a huge body of cold clear water ringed by rocks and studded with an alleged thirty-thousand islands. In 1615, Samuel de Champlain cupped his hands in Georgian Bay hoping to taste salt. Though not the first European to drink the water, he was in the vanguard. But he was probably the first to refer to Huron as *La mer douce*, the Sweetwater Sea.

Though I've never boated in Georgian Bay, I had recently driven around it with my family and was struck by the contradictions we saw. The bay is bounded by country that was wild only a few years ago and remains lovely, with the scent of wilderness lingering over it, but guerrilla developers had invaded. Fishing villages were being transformed into tourist destinations, and minor cases of sprawl were breaking out all over the place. Even the smallest towns longed for their Tim Hortons and Red Roof Inns. We were visiting in vacation season and the highways swarmed with American and Canadian tourists. On a Friday afternoon we got caught in a flash flood of weekenders roaring north from Toronto on Highways 400 and 69. Motels and restaurants charged top prices, and every private house, it seemed, hawked yard ornaments, birdhouses, windchimes, Authentic Indian Art, and firewood by the bundle. But the shores were gorgeous, and the water was clean and blue.

We spent a couple days at the village of Killarney, adjacent to Killarney Provincial Park. We hiked some of the shore and sea-kayaked a few miles of coastline. It's rugged and rocky, reminiscent of the north shore of Superior, but it lacks the open vastness of Superior and has more human marks upon it.

We noticed that the farther north we drove and the more remote the land became, the more likely we were to encounter small markers of stones erected to suggest human shape—a stack of flat rocks, with a wider one near the top to represent arms and capped with a round stone for a head. Perhaps they were modeled on the Inuit *Inukshuk,* which translates to "in the image of man." The Inuits relied on them to designate safe passage through the Arctic. But we saw them everywhere along the highways that parallel Georgian Bay and the North Channel, each placed prominently on outcroppings where you couldn't miss them as you drove past. They looked something like the copyrighted symbol used to promote the movie *Blair Witch Project.* The north shore of Superior has them too—not the shore itself, but the highway around it. My wife and I saw so many on Drummond Island, where we spent a day hiking on the fossil-rich limestone ledges of the northeast shore, that they became an irritation. Gail finally lost patience and began dismantling them, scattering the stones to a more natural disorder. But eventually she gave up; there were too many.

I thought they must be a kind of signature left by people uneasy in the wild. Humanizing nature is perhaps a comfort if you miss the company of others. No doubt they're fun to make, a project for the entire family. Search for suitable rocks, pile them like building blocks, step back and admire your work. But by the hundreds they begin to look like graffiti, which also is found in abundance along northern highways.

"Hendrix Lives!" says the graffiti.

"I was here!" says the Inukshuk.

We are strange and lonely creatures.

For dinner that evening Tim made fresh perch and shrimp, breaded and fried and served with garlic mashed potatoes and a Caesar salad. Tim was doing amazing things with the Dickenson diesel stove. Hajo claimed it was the best cooking stove in the world. He'd gotten on intimate terms with it while living aboard during early March, when the stove served for both cooking and heating. It was reliable, he insisted, and sturdy, and would work under any circumstances. Never mind that it stank. Or that a fine black soot settled on every surface of the cabin and seasoned our food with the faint flavor of diesel.

We sat down to the table, ravenous, but Tim announced that before we ate somebody had to tell a story. A tale for a meal, a reasonable barter. But we weren't in the mood to talk. We were tired and hungry. The food smelled delicious; we wanted to enjoy it in silence. But Tim insisted.

Finally, Hajo spoke. Ordinarily he would not sit still long enough to tell a story. At meals he usually took one bite and jumped to his feet to check the gauges. He took another bite and got up to examine the charts. One more bite and he climbed the companionway ladder to make sure everything topside was under control. At that pace a main course would take an hour to complete, but he never finished a course.

Now, though, he sat back in his seat, willing to talk. "Ever hear of Blue Domers?" he asked.

None of us had.

"It's my religion. Really, you've never heard of it? When I got out of college—well, first you have to understand that I went to Connecticut College and got a bachelor's degree in marine biology. This was the late sixties, peace and love were in the air, and my favorites, sex, drugs, and rock-'n'-roll. And this is important: I enrolled at Connecticut College the first year that men were allowed. Before that it was an all-girl's school. My freshman year there were something like eighteen hundred women enrolled and thirty-eight men. What was it Mark Twain said about not letting your schooling interfere with your education? That was good advice.

"So I got out of college and I'd had so much fun that all I wanted to do was have more of it. Some friends and I bought an old school bus and painted it in psychedelic colors and turned it into a hippy bus. We toured all over New England in it, picking up every hitchhiker we saw and getting high with them. In those days hitchhikers usually didn't have a dollar to help with gas but they always had a bag of grass on them. We tried to go to Woodstock but there was too much traffic, so we gave up and headed to Vermont, and after a while all my friends wandered off in their own directions and I was stuck alone on the bus with my ducks.

"Oh, yea, I had these three pet ducks that always rode in the back of the bus. The backseat was theirs. Every time we stopped for the night, I'd say, 'I gotta walk my ducks, man.' It never failed to get a laugh. I'd put a plank up to the back exit of the bus and my ducks would waddle down to the ground and follow me around wherever I went. They'd imprinted on me, like, what's that guy and his geese? Konrad Lorenz. He was a colleague of Niko Tinbergen's, one of my favorite authors, you should read his books about animal behavior. He was Dutch, like me.

"I'd walk down to the nearest stream or pond so my ducks could get a drink of water, and all these stoned hippies would see me walking with this string of ducks waddling along behind me quacking contentedly like I was their mother and the hippies would just freak out and laugh their asses off. They thought it was the funniest thing they'd ever seen. They didn't know if it was a hallucination or what.

"So after everybody splits, I'm in northern Vermont with my ducks and one day I run out of money so I drive up to this farmhouse in the middle of a Christmas tree plantation in the middle of nowhere and knock on the door. An old man answers and I ask him if he needs any help pruning the trees or anything, and he gives me a long look and checks out my hair and my bellbottoms and says he'll hire me if I give him ten minutes first to convert me to his religion.

" 'What's your religion?' I asked.

" 'Blue Domer,' he says.

" 'Never heard of it,' I said. And he says, 'Come with me,' and leads me around the house to a trail winding up a high hill in the woods. When we get to the bluff at the top, there's an incredible view of two or three counties of woods and hills with mountain ranges rising in every direction. The farmer sweeps his arms across the sky—across the whole big blue dome of it, you see—and says, 'This is it. This is my religion. Want to convert?'

"And that's how I became a Blue Domer."

The northeast shoulder of Michigan is the least developed and least visited shore of the Lower Peninsula. Tourist bureaus have tried gallantly to promote it as the "Sunrise Side," but so far the multitudes have stayed away. From out on Lake Huron the shore appeared low and heavily forested, interrupted here and there with small lakeside communities that from our distance remained nearly invisible. We spotted a few landmarks, like Old Presque Isle Lighthouse and Presque Isle Harbor. The French left their names everywhere on these shores. Similarly, their language permeates the language of the sea. Until Matt pointed it out to me, I didn't know for instance that "M'aidez"—"Help me"—became "Mayday," the universal distress call.

The first white people to see this shore were the procurers of souls and the traders of pelts. They came later here than they did in many other parts of the Great Lakes because of their fear of the Iroquois, who controlled the shores of Lake Ontario and Lake Erie and the portage around Niagara. Most of the eighteenth-century Jesuits and fur traders took the interior route, up the Ottawa River to Lake Nipissing and Georgian Bay. From there they coasted the North Channel to Mackinac Island or Sault Ste. Marie. The east coast of Michigan was bypassed until years later.

By the 1830s, after most of the beaver, marten, otter, and other fur-bearing animals had been trapped out of the Great Lakes watershed, attention turned to the other great resource of the region, its trees. Throughout the 1830s and '40s, agents for lumber barons skirted the shores of the lakes in schooners and steamer packets, then paddled their canoes up rivers to stake claims to some of North America's largest stands of white pine. Starting in Ontario, then moving west through northern Michigan, northern Wisconsin, and northwestern Minnesota, they found treasure growing from the earth. In Michigan, they ascended the Saginaw, Au Sable, Manistee, Muskegon, Grand, and other rivers until they reached the heart of the pine forests. Their reports described wonders. The pines reached 150 to 200 feet in height, with trunk diameters of 6 or 8 feet; a single tree could yield 6,000 board feet of prime lumber. You could walk for miles through the forest, beneath a canopy so dense no undergrowth could live there, over ground carpeted a foot thick with the accumulation of centuries of needle-drop. The trees towered overhead, and no branches grew for sixty or seventy feet above the ground.

By some accounts, the pines in Michigan extended in an unbroken forest from the Saginaw Valley to the Straits of Mackinac, and from the shore of Lake Michigan to the shore of Lake Huron—the entire northern half of the Mitten. Even if they grew in patches and thickets, as is more likely, they added up to thousands of square miles of virgin trees. At first the loggers were interested only in the big-ticket items, the mature white pines—"cork pines," they called them, because they floated so high in the water. Because of their strength and lightness and their straight, unblemished grain, they were in demand as ships' masts. In the many swamps and wetlands and wherever fires had burned were woodlots of less valuable aspens, maples, oaks, cedars, and spruces, making it necessary to search for the pines. A good way to find them was by ear. Every species sounds different in the wind. Pine sighs. You could hear a stand a half a mile away. To the timber cruisers, the sighing of those giants must have sounded like money.

The lumber cut in Michigan earned far more than all the gold found by all the miners in the California Gold Rush. From 1870 to 1880, Michigan led the nation in lumber production. The trees were felled, typically in winter, dragged through the snow by horse or oxen or mule, and stockpiled on high banks, or "rollaways," above the rivers. With spring's high water, the logs were rolled into the rivers and driven downstream to sawmills at the mouths. If no mills were there, the logs were chained into rafts up to a half-mile long and towed by tugs down the big lakes to mills in Chicago, Muskegon, Milwaukee, Detroit, and Buffalo.

It was a boom time of tremendous scale. Lining the final twelve miles of the Saginaw River were 112 sawmills working around the clock to produce, in 1882 alone, a billion board feet of lumber. Mills on a dozen other rivers were nearly as productive. For a couple decades the forests of the Great Lakes accounted for more than a third of the lumber produced in the nation. Some of it went east, to Albany, Philadelphia, Baltimore. Some was loaded on schooners and sent to Europe. Much of it was shipped to Chicago and distributed west to the wood-hungry settlers of the Great Plains. Scratch the rafters of old houses, barns, and churches in Kansas, Nebraska, Colorado, and Wyoming, and chances are good you'll find pine that once swayed in Great Lakes winds.

What came next is an old story. Some of the lumbermen were convinced that the giant trees would grow back faster than they could be cut and figured out ways to cut faster. Others had no illusions of sustainability—they just cut and ran. In their place came a second wave of lumbermen, then a third and fourth, each targeting smaller and less valuable trees, until even the aspens and jackpines were gone. As the trees were removed, the industry migrated west. By 1900, forty million acres in northern Michigan, Wisconsin, and Minnesota had been stripped clean. Instead of forests, there was the "cutover"—tens of thousands of square miles reduced to stumps, the soil eroding, the vistas lunar. The Great North Woods were gone. By 1920, trees were so scarce in the Great Lakes region that a "timber famine" was proclaimed and it became necessary to import wood products from the South and Northwest. Where the forests had stood there was now only slash—the discarded branches, the litter of the woods, heaps of it ten or twelve feet high carpeting hills and valleys as far as the eye could see. In the spring and summer, when it dried, the slash turned explosive. It was gasoline waiting for a match.

In the wake of the timbermen came land speculators and community boosters. They advertised for settlers, claiming rich land for the taking for enterprising go-getters who weren't afraid of a little work. "Cloverland," they called the former forests—fertile fields and pastures ripe, they said, for corn and cattle, once the slash was removed and the stumps grubbed out. (The stumps were a problem, they admitted, but one easily solved by fire, or a mule, or dynamite.) Homesteaders bought parcels cheap, then busted their backs to clear the stumps and manhandle their plows around the rocks and get a shack up before winter. But winter came early and stayed late this far north and the soil was terrible, mostly just "lean and hungry sand," in the words of P. S. Lovejoy, a University of Michigan forester who in the early 1920s became a forceful and eloquent advocate of reforestation around the northern lakes. Pockets of fertility existed there, but most of the land was fit only for pines. In the years after the Great War and before the Great Depression, nine out of ten northern Michigan farms failed. When farmers couldn't make taxes, millions of acres in Michigan, Wisconsin, and Minnesota reverted to government ownership. A lot of dreams died in that hungry soil.

A lot of them died in flames, also. Fires had been burning all along, of course. Settlers set them routinely to clear their land, and if the flames got out of hand and burned a few hundred or a few thousand acres more than intended, nobody worried about it. But then came the "holocaust fires."

The first and most horrific was centered around Peshtigo, Wisconsin, a lumber town located about forty miles north of the city of Green Bay, a few miles inland from the bay itself. Because the Peshtigo fire occurred on Sunday evening, October 8, 1871, the same night as the Great Fire in Chicago, it was largely overlooked by the national press. Even now the Peshtigo fire remains little known outside the Great Lakes region.

That year, the Midwest had suffered a severe drought. The previous winter had brought so little snow that logging operations had to be curtailed. Loggers required snow in order to haul massive logs by ox-drawn sled, and they needed heavy runoffs of spring meltwater to raise rivers enough to float the logs downstream. By spring that year, however, the rivers were too low to carry logs. The weather turned hot and dry, and stayed that way all spring, summer, and fall. Even the swamps dried up. Except for a brief sprinkle early in September, no rain fell on northern Wisconsin from July to October.

In spite of the drought that autumn, crews laying a railroad track between

Green Bay and Menominee, Michigan, burned brush and left the fires unattended. Farmers, loggers, and others were equally careless.

Peshtigo was a town of about two thousand residents, many of them employed in the sawmills or in a factory called the Peshtigo Company, which manufactured pails, tubs, broom handles, clothespins, and other woodenwares. Mountainous piles of sawdust surrounded the factory and the sawmills, and sidewalks all over town were made of boards that had dried and warped in the heat. The streets themselves were fuel for fire. For years sawdust had been sprinkled over them to keep the dust down.

All day October 8 the scent of smoke filled the air and the sky was hazy with it. That was not unusual. Small fires had been burning for months in the surrounding woods and slashings. But this day was so dry and hot—hotter than any autumn day anyone could remember—and the sky so dark with copper-colored smoke that people later recalled profound forebodings.

After sunset the sky turned crimson, as if reflecting distant flames. About nine or nine-thirty, as darkness descended, a wind rose from the southwest with a sound like locomotives bearing down on the town. People stood in the streets and watched to see what was coming.

It arrived with almost no warning. One minute there was smoke in the air and a distant roaring sound, the next the sky was raining gobs of fire. Roofs burst into flame, and the streets and sidewalks ignited. Then came a roar people said sounded like Niagara Falls or the world's worst hurricane, and hellfire slammed into Peshtigo. Some eyewitnesses saw "tornadoes of fire" sweep in from the forest; others saw walls of flame. The fire came faster than anyone could run, faster than horses and steam engines. Many people thought they were witnessing the end of the world.

Dozens died when they tried to take shelter in buildings that exploded into flame. Hundreds tried to escape by leaping into the Peshtigo River. Many survived, but some suffocated as the oxygen was sucked from the air, and others were swept downstream in the fast current and drowned. The fires were so hot that people submerged to their chins in the river felt their hair ignite. Logs floating in the millpond caught fire.

A blast of wind and fire struck the Peshtigo Company building and shot hundreds of flaming buckets and tubs through the air. People in the river had to duck underwater as the missiles struck around them in smoking, hissing explosions.

Within an hour, the town of Peshtigo was gone.

Farmers and loggers in the surrounding countryside died running or were incinerated in their homes or suffocated when they tried to take shelter in their wells. The remains of a family of eighteen were later found scattered around their farm. Some people committed suicide rather than die by the flames. A few men killed their families, then themselves. Some survived by covering themselves with soaked blankets and lying facedown in creeks, but whole communities were wiped off the earth. Across Green Bay, flames swept the length of the Door Peninsula, annihilating the towns of New Franken, Robinsonville, Brussels, and Little Sturgeon. All but seventeen of the seventy-seven inhabitants of Williamsonville perished. Nothing remains of that village today but a stone memorial.

In all, the Peshtigo fire killed twelve hundred people and burned a million acres of forest. It remains the greatest fire tragedy in U.S. history.

The same day, while Peshtigo and Chicago were burning, yet another fire broke out along the west shore of Michigan's Lower Peninsula. Fueled by winds off Lake Michigan, it swept across most of the northern half of the Mitten, engulfing two million acres and destroying the towns of Holland and Manistee and many smaller settlements. It ran down river valleys, engulfing pine forests that lined the Au Sable River, burning farms and sawmills along the Saginaw River. The fires didn't stop until they reached Lake Huron. In the town of White Rock, on the eastern shore of the Thumb, dozens of villagers took to the lake to avoid the flames that were destroying their homes. The fires were so intense that the flames were said to reach more than a hundred yards across the water. An estimated two hundred people burned to death in Michigan that day and fifteen thousand others lost their homes.

For the next few decades, fires were as regular as the seasons. The worst in Michigan was ten years after Peshtigo, in September 1881. It swept across a million acres in the Thumb of Michigan, burning two thousand square miles in a little more than five hours. Powered by winds of more than forty miles per hour, it rocketed through homesteads and villages. Again, people who sought shelter in their wells and root cellars were asphyxiated. Many took refuge in the Cass River, the region's major waterway. Survivors later claimed that they crouched to their chins in the water elbow-to-elbow with black bears, coyotes, bobcats, and deer, while flames jumped overhead and continued eastward. Nearly three hundred people died and another five thousand were left homeless.

A firestorm in 1894 engulfed Hinckly, Minnesota, and killed between four hundred and six hundred. In 1918, another tore through hundreds of thousands of acres in northeastern Minnesota, destroying the town of Cloquet and killing several hundred.

Those and other forest fires created an unexpected hazard on the lakes: thick smoke blowing across the water was responsible for dozens of ships colliding and running aground. Many of the 1,167 shipwrecks on the lakes in 1871, for instance, have been attributed to smoke. The worst loss of life may have occurred on October 18, 1871, ten days after the Chicago and Peshtigo fires, while much of northern Michigan still smoldered. That day the propeller-driven steamship *Coburn*, carrying forty passengers, a crew of thirty-five, and a cargo of wheat and flour, was downbound from Duluth to Buffalo. After passing Presque Isle Harbor in Lake Huron, she encountered thick smoke blowing off the mainland. The captain ordered the engine checked down, then steered the steamer into the wind, intending to hold steady until the smoke passed. But the wind veered suddenly from the northeast to the southwest and came up strong in a gale. High waves caught the ship broadside and tore her rudder away, leaving her rolling helplessly in the troughs. Her cargo shifted and she listed and began taking on water. Within a few minutes she sank. More than half the passengers and crew made it into lifeboats, but thirty-two drowned.

At midnight I climbed to the deck and sat in a chair on the poop deck. Hajo sat in another, one hand on the helm. The night was clear and cold, with a light wind. In the moonlight the breeze-ruffled surface of Lake Huron seemed to run like a river. The foresail caught the wind and made the rigging creak. By now I didn't notice the engine thrum unless I made an effort to listen.

To the west the dark shore of Michigan was dipping toward Saginaw Bay. This third largest bay on the lakes, exceeded in size only by Georgian and Green bays, is twenty-five miles wide and fifty long, 1,250 square miles in area, and shallow, most of it less than thirty feet deep. Concentrations of industries in Bay City, Saginaw, and Midland, and agricultural runoff from the fertile Saginaw Valley have raised phosphorous levels in the bay to among the highest in the lakes, making the water artificially fertile and very productive of fish. Unfortunately, the fish are contaminated with such high levels of PCBs and dioxins that state biologists recommend you never eat some species, including large walleye, the glamour fish of the bay.

Between the hourly bilge checks and log entries, Hajo and I talked. Hajo riffed on his past, elaborating on the life story he'd begun telling that first meeting at the dock in Traverse City. He was born in 1950 on the Indonesian island of Sumatra, on a 24,000-acre tobacco plantation that had been owned by three generations of his family. They grew "wrappers," the fine leaf used only for the outer layer of a cigar. Their house was a mansion maintained by a staff of thirty servants. Dutch colonialism had been very good to the Knuttels.

In many ways it was an idyllic childhood. Hajo's parents allowed him to roam freely about the plantation, his safety unquestioned. The Indonesians who worked for them and those living in the surrounding villages were so protective of children that it was like living in a nation of baby-sitters. Wherever Hajo went, adults watched over him, a necessary precaution on an island inhabited by tigers—Hajo remembers seeing several, including one on the porch of their house—and crocodiles and many venomous snakes.

Hajo's father ran the plantation efficiently, but it did not require his full attention. He was an accomplished amateur botanist and enthusiastic about art, music, and literature. Hajo's mother kept busy managing the household and servants, and enjoyed entertaining guests. Nearby was the summer home of Hajo's grandfather, who often hosted the queen and prince of Holland when they visited Sumatra.

Though Hajo had an older brother and sister, he chose to spend most of his time with his pet orangutan, a female named Noirtje (pronounced

"NOR-cha"). They were the same age, but Noirtje grew to be much larger than Hajo. "She was stronger than I," Hajo said, "but I was smarter."

They became inseparable friends. Every day they wandered the plantation together, looking for adventure. Afternoons they loved to sit in the sand in the shade of a large tree in the backyard and play with Tinkertoys. But Noirtje often became restless. Through the open door of the laundry room she would watch washerwomen gossiping as they ironed and folded clothes. Hajo would follow her gaze and warn, "Don't do it, Noirtje." But the orangutan would sidle casually toward the door, then suddenly bolt for it—Hajo shouting, "No! No! Don't do it!"—and charge inside the laundry room, braying wildly, grabbing clothes and throwing them into the air. The women always screamed and ran from the house and refused to work any more that day.

Hajo's mother would be furious. She would scold the orangutan and order Hajo to put her in her cage. This was stern punishment. Noirtje hated confinement. If anyone but Hajo tried to punish her, she would climb a tree and refuse to come down.

"I would agree to cage Noirtje," Hajo said, "but only on the condition that my mother give me two pieces of penny candy first. I would then explain to Noirtje that one of the candies was for her, but she must agree to go into her cage before I would give it to her. It was the only way she could be coerced into it. So I would hand her the piece of candy and she would look very downcast, but she would honor the deal and allow me to lead her to the cage. Inside, she would sit in a corner and very carefully unwrap the candy with her fingers. She would hold it and look at it for a long time and sniff it lovingly. After a while she would touch the tip of her tongue to it and roll her eyes with pleasure. She would take another dainty lick and roll her eyes again. In this way she could make one piece of candy last all day."

As all pet stories must, this one ended sadly. Sumatra had been under Dutch control for centuries. Though the nation became a republic and gained independence in 1949, the Dutch did not lose effective control until 1958, when dissident army officers rebelled. The new authoritarian regime seized the assets of foreigners and kicked them out.

The Knuttels were driven from the country with only what they could carry. Noirtje had to be left behind. Hajo remembers riding in a jeep away from the plantation in the middle of the night, while their former employees lined the road, many of them weeping. Even under such terrible circumstances, their home and possessions taken from them, their lives in danger, Hajo's father bothered to jam on the brakes of the jeep and stop when a snake crossed the road in front of them. He got out to investigate—he wanted to know what species it was. At the airport he bartered his gold watch for plane tickets for his family.

They applied for asylum in the United States and moved to Connecticut, where they were sponsored by a Jewish businessman. In a twist worthy of Hollywood, Hajo's father had helped this same man and his family escape occupied Holland during World War II. Now the man returned the favor. He loaned Hajo's father enough money to buy eighty-five acres of good farmland along the Connecticut River. There the family began raising azaleas and rhododendrons and selling them wholesale. Within a few years Hajo's father was known for his hybrids, some of them still popular today. In time the family prospered again.

But there was more. Five years after leaving Sumatra, when Hajo was thirteen, he traveled alone to Holland to visit his grandmother in Rotterdam. One day they decided to go to the zoo. While walking among the displays Hajo saw an orangutan in a raised cage of the sort now condemned as inhumane, with bars on all sides and a chain fence around it to keep people out of reach. The ape stood with its back to him, but Hajo knew immediately that it was Noirtje. He called her name. She turned. She recognized him. Hajo ducked beneath the fence and put his hands between the bars, and she placed her palms against his palms and they looked into each other's eyes.

A zookeeper saw Hajo and rushed forward shouting. The grandmother intercepted him. She raised her cane in the air and threatened to beat him with it if he dared lay a hand on her grandson. The keeper ran for help. In a few moments he returned with the superintendent of the zoo.

Hajo and his grandmother told their story. The superintendent explained in turn that Noirtje had been sold to the zoo shortly after the revolution, by the man Hajo's father employed as a gamekeeper on the plantation. His job had been to keep monkeys, snakes, and other nuisance animals under check. He often earned extra money by selling them to zoos.

The superintendent gave Hajo permission to enter the cage with Noirtje. The next few afternoons he joined her there, the two of them sitting on the floor of the cage, holding hands, talking quietly. But soon Hajo had to return home to America. He never saw Noirtje again.

"Someday I want to return to Sumatra," Hajo said. "But I'm sure it will be very different there. And probably very sad for me."

Far in the distance the Michigan shore glittered with faint lights, all the good people asleep in their houses. We were cold, bundled in our sweaters and coats, our foul-weather jackets on for their slight insulation and their tightly drawn hoods. We scanned the lake for freighters and thought about the strange bounty of the earth.

I told Hajo his story could have been written by Kipling. I wasn't entirely sure whether to believe it.

"Almost everything in my life has gone wrong," Hajo said. "Isn't it wonderful?"

Chapter 9

ST. CLAIR RIVER/DETROIT RIVER

Low Water in the Lakes ◆ To the St. Clair ◆ Freighters and Lake Carriers ◆
Run Aground ◆ La Salle and the *Griffin* ◆ We Dock in Detroit

It took all night to pass Lake Huron's Saginaw Bay, a crossing feared for good reason in the days of the voyageurs. Many travelers in small boats have been forced to take shelter on tiny Charity Island, and many others never reached shore. The bay is shallow, less than twenty feet deep most of the way across, and it gets treacherous in a wind.

At dawn we could see the Thumb of Michigan to the west. Inland is low and fertile farm country (Huron County, encompassing all the tip of the Thumb, grows more dry beans than any other county in America), but the shore itself is made of limestone and has been the death of hundreds of vessels.

Then came the narrowing southern end of Huron, where the water of all three upper lakes funnels into the St. Clair River. Matt, Harold, and Tim came on deck to relieve us, but Hajo and I stayed up, not wanting to miss anything.

While still several miles from the river at the end of the lake we became aware of a large structure near shore. As we watched, optical distortion transformed it into various grotesque shapes. We took turns examining it through the binoculars, finally agreeing that it must be a vessel. As we got closer, we saw that it was a freighter grounded on the flats east of the river. It was one of several ships that ran aground in 1999 and the spring of 2000. Water levels were approaching all-time lows, and many channels and connecting rivers that were normally deep enough for navigation were becoming questionable. Lake vessels were forced to load smaller cargoes in order to negotiate the shallow places.

I remember the early 1960s, when the lakes were nearly as low as now, and, more vividly, the 1970s, when they were so high that homes along the shores collapsed into the lakes as waves ate the land from under them. Such fluctuations have always been regarded as a consequence of natural cycles of rain and snowfall, often associated with El Niño and La Niña weather patterns originating in the Pacific Ocean. The geologic record shows periodic swings in water level going back thousands of years. Naturally, the question of global warming has been raised. Most scientists agree that the surface temperature of the planet is rising, but there is less agreement over the cause of it and what it could mean to a hydrologic system as complex as the Great Lakes. One recent computer model projected a period of drought and heat continuing through the twenty-first century, resulting in even lower water levels. Another predicted more heat and more precipitation, resulting in the Great Lakes staying at the same level or even rising a foot or so above average.

Whatever the prognosis, it is certain that since 1997 precipitation has been declining and lake levels dropping. Warmer weather had to be a factor. Six of the hottest years ever recorded occurred in the 1990s. Every winter, smaller portions of the northern lakes and their bays were freezing, allowing evaporation to occur during all twelve months of the year instead of eight or nine. By the summer of 2001 the lakes were three feet below normal, approaching their lowest levels since 1860, when records first began being kept, but a year later, levels had begun to climb, and the trend seemed to be reversing.

Everywhere around the lakes marinas and harbors were being dredged in an effort to keep them open. The six-billion-dollar-a-year Great Lakes shipping industry was hurt badly. A spokesman for the Lake Carriers' Association had recently announced that the low water forced freighters to reduce their cargoes two or three thousand tons below what they carried a year earlier, amounting to a loss in revenue of about $100,000 per vessel per voyage. The problem was compounded by severe declines in U.S. steel production. Every ton of steel requires two tons of raw materials—iron ore, coal, and fluxstone—and the bulk of the domestic supply is shipped on the Great Lakes. One shipping company president summed it up like this: "When the steel industry gets a cold, shipping on the Great Lakes gets pneumonia." In 2001, ten of the seventy vessels in the Great Lakes fleet were berthed part of the season, waiting for higher water and better economic times.

Low water might also prove catastrophic for life in the lakes. Warmer water and shrinking shoreline habitats would kill wetlands and reduce spawning and nesting of indigenous species. It could also open the lakes to invasion by animals and plants that thus far had been repelled by cold water. The algae that form the foundation of the food web might not prosper in lower, warmer water—or might prosper too well.

Almost certainly the water will rise again. Rain and snow will increase, rivers will fill to their banks, the lakes will come up. But maybe not as high as before. Maybe from now on every low in the cycle will pull the average down.

Seeing a ship the size of a hotel stranded and listing a mile out of the shipping channel was disorienting. The vessel had lost its way in fog, perhaps. In the hierarchy of maritime disasters, running aground is not as serious as collision or sinking. But it is a big blunder. It can end a captain's career. Hajo slowed the *Malabar* and kept careful watch of the sonar depth finder. We stayed near the center of the shipping channel, safely between the red and green buoys, determined to take no chances.

In the early evening, the sun dipping, we entered the St. Clair River and felt the pull of gravity carrying us along. We passed beneath the Blue Water Bridge linking Michigan and Ontario. After three days and nights on the water we were anxious for showers and an uninterrupted night's sleep—there would be no need to stand night watch while docked. Hajo had radioed ahead to reserve space at a marina somewhere near the bridge. *Malabar* could not tie up just anywhere; we needed 8.5 feet of water and enough dock to accommodate 105 feet of length. The man Hajo talked with promised to hold a spot for us.

But the marina we presumed was in Port Huron, on the St. Clair River, turned out to be many miles downstream in Detroit, on the Detroit River. The confusion was about bridges. The man at the marina had described his location near the Ambassador Bridge, which crosses the Detroit River between Detroit and Windsor, Ontario. Hajo and Harold thought he meant the Blue Water Bridge, at Port Huron and Sarnia. It was an understandable mistake, but disappointing.

We passed downtown Port Huron and considered swinging into the mouth of the Black River, a small tributary where the Port Huron–to–Mackinac Race commences every year. But when we hailed several sailboats near the mouth and asked if the river was deep enough for a boat our size, everyone said no, definitely not.

We could find no place to dock. The American side of the river near Port Huron is lined mostly with homes, each attended by a sailboat or motor yacht moored to a small dock in front. The Canadian side is lined with some of the most odious industrial jumbles on the lakes: wharves and loading docks and refineries crowded with petroleum storage tanks, with belching stacks and steaming pipes. Canals branching off the river are filled with acid-scented water the color of rust. In some of the canals are the hulks of freighters apparently abandoned there. Some were in the process of being dismantled, plates of rusting steel stripped from their hulls.

With night approaching and nowhere to tie up, our only choice was to push on. We decided to descend the river to Lake St. Clair, where we could leave the shipping channel and find an anchorage in protected water.

The large and powerful St. Clair River was unfamiliar to us, and darkness was falling quickly. The river is a half-mile wide much of its length, but the charts show only nine hundred feet of channel deep enough for navigation. Nine hundred feet seems like a lot until you meet a lake carrier, then it shrinks alarmingly. The current is strong and deceptive—turbulent below, smooth on the surface. It carried us along at eight knots, faster than we wanted to go.

A retirement-aged man in a small aluminum fishing boat trolled beside us. He was using a technique designed to get around the challenges of deep water and strong current. Instead of a rod and reel, he used a hand-cranked wheel mounted to the transom of his boat. Around the wheel was wire line attached to a lead ball weighing several pounds. To the wire above the weight he had attached short lengths of monofilament line, each terminating with a fishing lure. To use this rig, he unwound the wheel, lowering the weight and lures toward bottom, then adjusted the speed of his motor until the lures agitated properly. When a fish struck, he reeled the entire rig to the surface.

He hailed us and pulled up alongside. He asked where we were from and where we were going. We told him, and he grinned. Need another crewmember? I'd go with you if I could. Want some fish? And he reached into a chest cooler in the bottom of his boat, pulled out a three-pound walleye, and tossed it on our deck. He tossed two others as well.

Tim went below and returned with a copy of his *Buck Wilder's Small Fry Fishing Guide*. He asked the angler if he wanted it autographed.

"I already own that book," the man said, not as surprised as you might think. "But I'll take a copy for my granddaughter, if you don't mind."

I filleted the fish while Tim lit the charcoal in the Weber, and twenty minutes later we were eating grilled walleye and fried potatoes. Bounty of the earth.

Then it was dark, and we got busy trying to decipher the buoy lights in the channel. Most of its length the river is lined with houses and roads, creating a confusion of illumination. The reds and greens of the navigation buoys get lost in the general blinking and flashing of taillights and yard lights. It required four sets of eyes and much discussion to decipher them.

Harold and I took the bow. Whenever we spotted what we thought was a buoy or another vessel, we would hurry to the stern and report to Hajo. "There's one, flashing red, dead ahead," we might say.

"Where?"

"Straight ahead. Now just to port of bow."

"Is it a buoy or shore light?"

"Buoy, I think. I'm sure of it. Pretty sure."

"Flashing?"

"Every four seconds."

"Is that it?"

"Just to port."

"Okay, I see it. Where's the green? Anyone see the green?"

We motored thirty miles down the river that night and our confusion increased every mile of the way. It got worse near the end of the river, where it divides into channels at the delta before Lake St. Clair. Matt stayed aft with Hajo and pawed through navigation charts by flashlight. Every few minutes Harold or I would run back with news.

Harold: "Blinking red on the port side, Captain, looks like a buoy."

Me: "Uh, there's something up there, I can't really tell what. Might be a parked car on shore or maybe a small boat at a dock."

At some point we became aware of a large vessel bearing down on us from our stern. It was one of the giant bulk carriers, four or five stories high, with the wheelhouse and cabins high in the bow, the engine room aft, and the hold with its rows of cargo hatches stretching between. The design is unique to the lakes and dates back to the 1880s, but the prototypes were only a couple hundred feet long. This was a modern vessel, much larger than that. Hajo was already anxious. Eight hundred feet of bulk carrier bearing down on him made it worse.

The ship kept coming. It seemed as big as a football stadium. In the night, with only shore lights and buoys for perspective, it was impossible to tell if it was a hundred yards from us or a thousand. It disappeared behind a bend in the river, then reemerged in a monumental glide—a city block of buildings swinging around the bend, silent and imposing, moving at first only in relation to our own progress, then under its own power. Faster than we expected. And definitely gaining on us. What were its captain's intentions? Did he plan to pass us? To nudge us out of the way? Did he even see us? Did he give a damn?

Matt swung a spotlight in the air, careful not to aim directly at the ship. The beam of light stroked from shore to the tops of our masts to the water beside us and to shore again. Certainly we were seen. There would be a man on watch in the bow. He would have spotted us long ago and notified the helmsman in the aft.

We entered marshlands and knew we must be getting near the mouth of the river. Matt ran forward with his spotlight and scanned walls of marsh-grasses, a precise circle of grass coming into vivid focus wherever he aimed. He said nothing to Harold and me and hurried back to Hajo.

Then the ship was only a hundred feet away, and gaining, and we could hear the massive engines, deep and powerful, like the dynamos at a generating plant. The bow towered above us, closing the distance. It was enormous. Ahead of it was a bow wave, frothing white at the top, high as a house.

On the right side of the river was a green buoy mounted on a concrete pedestal. It marked the mouth of a channel or bay. To me, it looked like the channel led into a bayou. It was difficult to tell in the dark, of course, but shore lights on the far side glittered across a surface that had the dimpled look of shallows, with perhaps aquatic weeds growing to the top. I'd fished enough at night to recognize that kind of water. The surface conceals most of what lies beneath but it offers hints. We passed the buoy, the black river swirling in its lee.

The *Malabar* swung sharply to starboard.

My first reaction was relief. I thought Hajo was pulling over to allow the ship to pass on our port side. It seemed the logical thing to do.

But no. Later Hajo would explain that he was afraid the big vessel passing in such close confines would suck our boat into its stern. Matt, poring over the charts (we would learn), interpreted the opening behind the buoy as a navigable channel. He was not mistaken. It had been navigable in shipping seasons past, when water levels were higher and the secondary channels were dredged. But dredging had ended years ago and silt had replaced water. The channel was suitable now for small fishing craft and pleasure boats, but not deep-draft vessels like ours.

We turned hard to starboard and kept turning. Realizing what was happening, Harold and I both shouted, *"No!"* But too late. We turned ninety degrees from the river, aimed now toward the shore lights across the water in Algonac, and charged forward into the old channel. There was trembling resistance beneath the hull and the bow began to rise. We slid to a sickening, horrible, forward-lurching halt.

Aground.

In the euphemistic jargon of the sea, we had "touched bottom." Some touch. We were firmly aground, our hull buried deep in mud. The lake carrier slid past with the mass of a skyscraper being towed down an avenue. The stern rumbled by, its propeller groaning in rhythm, and we rocked in its wake. Hajo threw the engine into reverse and accelerated it to a howl, but the *Malabar* didn't move.

Matt and I put on life preservers and lowered the yawlboat to the water. We climbed down into it and pull-started the outboard. Slowly we circled the *Malabar,* checking depths with a pole. Bottom was six feet deep near the bow, eight feet at the stern. I held a spotlight on the bow of the *Malabar* and we bumped against it with the yawlboat. Matt accelerated the outboard to full throttle, while Hajo reversed the engine of the larger boat and accelerated. There was much roaring and thrashing. Mud and weeds and a stench of decay came to the surface. But the *Malabar* did not budge. We tried pushing sideways against the bow, but still she would not budge.

Matt and I tied the small boat to the side of the *Malabar* and climbed on deck. The air was full of tension. We decided to wait for another lake carrier and try riding its wake. It was a plan, better than radioing for help. Hajo made it clear that we must avoid involving the Coast Guard. Reams of paperwork would need to be filled out, and Steve Pagels would be notified. There would be an inquest. If Hajo was found negligent, he might lose his captain's license, unjust as that seemed. This was not Hajo's fault. I blamed the captain of the freighter. He was probably laughing at us at that very moment. I imagined him as the kind of man who would radio the crews of other lake carriers and tell them about our misfortune. "Check out the schooner on the mudbank," he might say. "Oops. My bad."

Nobody talked. We were sick at heart, tired and worried. Matt and Hajo were furious with themselves and each other. Matt would later admit he had been wrong in advising Hajo to turn into the channel. When he ran forward with the spotlight and saw no opening in the marsh, he thought the only route open to us was to the right. Hajo would say only that every decision was the captain's responsibility—the buck stopped with him.

A ship came downstream, then another, but though their wakes made *Malabar* rock on her keel as Hajo reversed the engine, she did not budge.

It occurred to me that an upbound vessel, going against the current, might push a larger wake. Matt said, "I doubt it, but it's worth a try." Hajo insisted we all get some sleep and wait until morning, when we would try an anchor maneuver, ferrying the anchor in the yawlboat to the river and trying to winch the *Malabar* to it.

None of us slept much. Hajo kept the engine idling all night. Each time he heard the rumble of an approaching ship, he ran up top to the helm and tried reversing as the wake of the ship washed beneath us. Only two or three boats passed, all downbound. Traffic on the St. Clair was much heavier a decade or two ago. For most of the twentieth century more tons of cargo were shipped every year on the Great Lakes than in all the American ports on the Atlantic and Pacific coasts combined. But these are tough times for the Great Lakes shipping industry. In the 1960s, a bulk carrier or freighter would have passed us every twelve minutes on average. Now it's a busy day if twenty ships use the St. Clair River in twenty-four hours.

At dawn, with gray light showing in the cattail marshes, three bulk carriers in close succession churned upstream from Lake St. Clair. In my bunk I heard our engine accelerate until it roared. The boat shifted slightly. I thought I had imagined it, but again there was a shift. We were backing! I jumped from my bunk a moment before Harold leaned down the stairs and shouted: "All hands on deck! All hands on deck!"

I hurried up top. Matt had gone over the side into the yawlboat, where we had left it tied, and was struggling to start the outboard. The *Malabar* was backing quickly into the river, the current rushing beneath us. The stern of the yawlboat pushed a wave that threatened to bury the transom. Matt stood, yanking furiously on the starter cord. Hajo struggled with *Malabar*'s helm and throttle, trying to back the boat upriver so the bow would swing downstream. But the third lake carrier was plowing past, so close that we were in danger of ramming it with our stern. Harold and Tim tried to release the mooring lines attaching the yawlboat to the side of *Malabar*, but the knots were too tight. Hajo shouted at Matt to climb out of the boat. Matt didn't hear, or ignored him. He yanked the starter, yanked again, adjusted the choke, yanked. Hajo shouted at him again. Matt looked up and began to protest.

Hajo bellowed, *"Get out of that boat now!"*

Matt obeyed, the captain's authority final. He scrambled over the side and into the *Malabar*. We released the lines at last and hand-lined the yawlboat to the stern, which was what Hajo wanted done in the first place. Matt climbed down into it, attached the hawsers to the bow and stern, and we winched the boat until it was suspended safely six feet above the water. We secured it there.

The third ship slid past, a few crewmen watching impassively from above. We swung around in the current, Hajo shifted the engine to forward, and we bucked over the big ship's wake and headed downstream.

We motored down the last few miles of river, staying in the South Channel—easily seen now as the navigable route—past Harsens Island and the St. Clair Flats Wildlife Area. Harsens Island is woven with canals, dikes, levees, and bridges. It's a place for long views, where the highest elevations are the roadbeds a few feet above the water. A hundred years ago the island was promoted as the "Venice of America," and was a nationally famous tourist destination. It's quieter today, though popular during duck season.

We passed marshy Seaway Island and entered Lake St. Clair at the St. Clair Flats. This is the most extensive freshwater delta in America. Channels and canals are surrounded by miles of waving reeds and bullrushes, forming remarkably fertile habitat for wildlife. Before dredging in the nineteenth century deepened the rivermouth, the St. Clair Flats were notoriously difficult to negotiate and many boats got stuck here. For us, there was little consolation in this knowledge. Still, I was interested to see the place where the *Griffin* ran aground during her maiden voyage in 1679. The *Griffin* was the first sailing vessel to travel the Great Lakes above Lake Ontario. Her history is strange in the extreme.

Few people made a greater impact on the Old Northwest than Réné-Robert Cavelier, Sieur de la Salle. He was born in 1643 in Rouen, France, the son of a wealthy and influential merchant, and grew into a proud and ambitious young man who cared less for pleasure than for accomplishment.

Like many other restless young men of the seventeenth century, La Salle left France and sailed for Canada. He arrived in Montreal in 1666, at the age of twenty-three, and found it a bustling and dangerous place—the most dangerous in Canada, some said. Warriors of the Five Nations of the Iroquois, the most feared of the eastern tribes, roamed the outskirts of the settlement. La Salle solicited and received a large grant of land about eight miles from Montreal, above a rapids in the St. Lawrence that would in a few years become known as *La Chine*—"On to China"—a sarcastic reference to La Salle's dream of a westward passage to Asia. The tract granted to La Salle was pure wilderness, exposed to attack from all sides. Immediately he began construction of a pallisaded village. He soon had leased acreage to settlers and built an outpost for the fur trade. At the same time he set to work learning to speak Iroquois and seven or eight other native languages and dialects.

La Salle would likely have become a wealthy man had he tended his wilderness kingdom and skimmed a share of the furs flooding down from the interior. But the life of a merchant was too tame for him. He dreamed of finding an inland route to the Pacific and the glory that would come with it. His dream would cost him his life.

In January 1679, he put thirty men to work carrying lumber and supplies up the steep twelve-mile portage around Niagara Falls. They cleared space for a small boatyard near the mouth of the Niagara River, at what is now the Buffalo waterfront, and the shipbuilders and their assistants went to work. The vessel they built was sixty feet long, of forty-five tons burden, with a high poop deck, two square sails, a griffin for a figurehead (the head and wings of an eagle, body of a lion), and five small cannons. They raised her sails and towed her with ropes along the shore until she was free of the current of the Niagara River at the entrance to Lake Erie. On August 7, with a crew of thirty-four men, they set off across the lake.

La Salle had a gift for surrounding himself with colorful personalities. His second in command, though not along on the first voyage of *Griffin* (he had gone ahead to Mackinac in a canoe to see to affairs there), was Henri de Tonty, an Italian count and military officer who came to the New World seeking adventure. He had lost his right hand to a grenade in the Sicilian wars and now wore an iron prosthetic. At a gathering of Indians and Frenchmen he was challenged to a fight and busted his opponent's skull open with his fist. The rest of his life he was held in awe by the Indians, who called him "Iron Hand." He would prove to be loyal, brave, and reliable, three qualities that were rare among La Salle's companions.

In attendance on the boat was Louis Hennepin, a Flemish Récollet friar who was equal parts liar and braggart, yet remained steadfast and periodically courageous during several years of travels with La Salle. He filled his journals and letters with exaggerations, usually to pad his own accomplishments. In the later editions of his memoirs he assures readers of his honesty, writing: "I here protest to you, before God, that my narrative is faithful and sincere, and that you may believe every thing related in it," then proceeds to lie through his teeth. He claimed, for instance, that he was the first white man to descend the Mississippi River to its mouth. The first had been La Salle, but he was safely dead by the time Hennepin revised his memoirs, and Hennepin must have thought no one would challenge his claims. Historians are quite convinced that Hennepin never saw the Mississippi below present-day St. Louis.

They sailed three days across Erie, then up the Detroit River to the lake they named Sainte Claire. At the St. Clair Flats they had difficulty finding a channel deep enough for the *Griffin,* but finally got through. In the St. Clair River below Lake Huron, they met current too strong to sail against, even with the help of a brisk south wind, and sent twelve men ashore with a tow rope to haul the boat the rest of the way up the river.

On Lake Huron they at first had good sailing, but a gale rose and they were battered by steep fierce waves unlike any they had seen. The pilot of the ship, a Dane named Lucas, was possibly a secret enemy to La Salle, who had many—New France boiled with resentments, jealousies, and political intrigue, and La Salle was too ambitious to be loved. In that storm on Lake Huron, the pilot alone among the crew refused to fall to his knees and pray for deliverance. Instead, according to Hennepin, ". . . he did nothing all that while but curse and swear against M. la Salle, who, as he said, had brought him thither to make him perish in a nasty Lake, and lose the Glory he had acquired by his long and happy Navigations on the Oceans of the World."

But the prayers of the others must have been heard. The storm abated and *Griffin* sailed on to Mackinac, then to Lake Michigan. At Washington Island, near the entrance to Green Bay, the crew collected thousands of furs, enough to fill the hold of their ship. Then La Salle made an unwise decision. He gave his possibly traitorous pilot charge of the *Griffin*, with orders to sail her to Niagara, unload the furs to be delivered to Montreal, and return to Green Bay. La Salle stayed behind with fourteen men in four canoes, and on September 18, 1679, with the wind light from the west, watched his ship depart.

It was the last he would see of her. That night a storm arose and blew for four days. A party of Indians later claimed to have seen the *Griffin* anchored near the top of Lake Michigan. They said they advised her crew to sail close to the coast and not attempt to enter the open lake in such a wind. But the pilot, wrote Hennepin, "was dissatisfied, and would steer as he pleased, without hearkening the Advice of the Savages, who, generally speaking, have more sense than the Europeans think at first . . ." Nobody knows what happened. There were rumors that the ship was boarded by Ottawas or Pottawattamies, who murdered the crew and burned the ship. La Salle himself became convinced that the pilot, whom he had suspected was capable of treachery, scuttled the ship and stole the furs.

Though conclusive evidence of the *Griffin*'s fate has never been found, there have been tantalizing suggestions. In 1890, a lighthouse keeper on Lake Huron's Manitoulin Island found four skeletons in a cave. Nearby he found two more skeletons, a pocketwatch and some seventeenth-century coins, and part of a ship's hull. A bolt removed from the wood was determined to be similar to those used in the days of the *Griffin*.

In 1955 a retired fisherman named Orrie Vail and a reporter for the *Toronto Telegram* recovered timbers and ironwork from a wreck Vail had found in a snug inner harbor on an island near Tobermory, at the point of Georgian Bay's Bruce Peninsula. They became convinced that they had found the resting place of the *Griffin*. In piecing together admittedly scant evidence, they conjectured that Lucas the Dane had talked a party of Indians into helping him hide the ship and dispose of the furs. Perhaps the Indians guided the ship into the tiny cove—it seemed unlikely that a vessel that size could have sailed in on its own—then murdered the crew and took the cargo.

La Salle went on to establish forts at the mouth of the St. Joseph River in southern Lake Michigan and on the Illinois River not far from present-day Chicago. He pushed westward in canoes, always looking for a route to the Pacific, and with Tonty and a small party of men descended the Mississippi to its mouth in 1682, claiming the lower river and surrounding lands for France and naming it Louisiana. Two years later, after returning to France in an effort to win political support, he sailed with four ships for the Gulf of Mexico, planning to build a fort at the mouth of the Mississippi and take charge of all the land between there and Lake Michigan. But he overshot the mark and landed in Texas instead, where his luck ran out. After several attempts to march overland to the Mississippi, some of his men, never loyal to begin with, became mutinous and murdered La Salle.

On the *Malabar* the mood had improved. Nobody was quite singing with happiness, but the relief was palpable. We had escaped a grounding, and nobody—not Steve Pagels and certainly not the Coast Guard—needed to know how close to disaster we had come. Ahead of us two lines of buoys aimed across Lake St. Clair toward Detroit. Lake St. Clair is big but so shallow on average that a boat our size must cross it between the buoys strung across its center. The lake is too long to see across from end to end, or was the hazy morning we crossed. From the center we could glimpse the shores on either side, but they were very low in the distance.

Near the end of the lake the shores drew close and we passed the communities of Grosse Pointe Shores, Grosse Pointe Farms, Grosse Pointe, and Grosse Pointe Park. Large, expensive houses lined the shore; canals went away inland and we could see yachts docked in them. I took bow watch as we entered the Detroit River. The river is similar to the St. Clair in size and character, but its shores are more heavily developed. We passed Belle Isle, a city park designed in part by Frederick Law Olmsted, with picnic grounds, fishing piers, zoo, aquarium, and the Dossin Great Lakes Museum. Detroit's skyline rose above the river downstream.

Tim came forward. We were alone in the bow.

"Let's talk," he said.

"Sure."

"We can get off the boat here."

I didn't know what he meant.

"We can get off in Detroit. I'll make a phone call and in four hours someone will pick us up at the marina and take us home. We can eat in a restaurant tonight. Have showers. Sleep in our own beds. I don't know about you, but this isn't what I expected. It isn't fun. I've had enough."

I was disappointed. I had assumed everyone was having as fine a time as I. But now I recognized signs of his discontent. Back in Traverse City, Tim had come to the dock nearly every day to watch Hajo and Matt complete the repair job. "This boat going to make it?" he would ask, grinning, and they assured him it would, throwing in "we hope," and "only one way to find out," and "good thing we've got lots of life jackets," just to be funny. But it fueled doubts that would nag Tim the entire trip, and finally make him quit the ship in New Haven, before entering the open ocean. There were other factors, too. His dislike for Matt. The increasingly unpleasant odors from the head next to his cabin. The low doorways where he kept striking his forehead. The general filth and low comfort level of life on a boat. But above all, the concrete hull. He would later admit he had been terrified much of the way through the lakes and on Long Island Sound. He always expected the next wave to break the boat in half.

"I'll quit if you quit," he said.

But I didn't share his feelings. "Sorry. I'm staying."

"You're sure?"

"Yes."

"Okay, then so am I."

"Good," I said, and I meant it. We needed all the help we could get. And Tim was good company, always cheerful, always ready with a joke or a story, and an excellent cook. Without him I knew we would be eating standard bachelor fare and the galley would soon overflow with dirty dishes.

Almost downtown, with the sleek cylindrical towers of the Renaissance Center rising high above the river not far downstream, we moored at Gregory's Fuel Dock. We topped off the *Malabar*'s below-deck fuel tank, then the two 55-gallon drums we had emptied on the trip down. We had burned diesel at the rate of a gallon every four miles.

I stayed on the dock and threw off the dock lines. The current swung the bow into the river. Hajo motored the *Malabar* upstream a hundred yards, turned her, came back with the current, aimed for the mooring dock, and accelerated. To make the entrance, he had to adjust intuitively for the four-mile-per-hour current. His judgment was perfect. The big boat passed through the marina entrance with only a dozen feet to spare on each side.

Meanwhile, I was in the yawlboat with the little Mercury outboard running. I shot ahead into the marina to a spot near the back wide enough to turn the *Malabar* around. Moored at docks on the perimeter were pleasure boats and yachts. At a signal from Hajo, I bumped the nose of the yawlboat against the bow of the bigger boat and twisted the accelerator to full throttle. I was a sheepdog nudging a cow toward the barn. Hajo maneuvered not by steering but by shifting between forward and reverse, the offset propeller turning the boat. The yawl churned the water mightily and slowly pushed the bow of the *Malabar* around. When it pointed toward the Detroit River, I shot ahead to the dock assigned to us, tied off the yawlboat, and stood ready to catch the mooring lines.

Hajo brought the *Malabar* forward. She drifted slow as a zeppelin and came to a stop at the precise moment that the bow and stern touched the dock. Perfect. "I don't control the boat," Hajo would later insist. "The boat goes where she wants to go. I'm just there to help." Harold handed the swing line to me, then dock lines one, three, and four, and I looped them over cleats on the dock. Harold, Matt, and Tim drew the lines tight and belayed.

We would stay in the marina for two days and nights, resting and recreating. That first day Matt's fiancée arrived, driven to Detroit from Baltimore by her father, Walt. Lisa Meier was lovely and spirited and would brighten our trip for the next week. She was thirty-one years old, had grown up in Wilmington, Delaware, and been sailing since the age of seven, first on inland lakes, then on Chesapeake Bay and the open Atlantic. After earning her BA in history at Wesleyan and her master's in social work at

Bryn Mawr, she moved to northeast Maryland and went to work for Child Protective Services for the state of Maryland. She had met Matt three years earlier in Baltimore, after she volunteered on a whim to be a deckhand on the schooner *Nickel*, where Matt was first mate. They planned to be married in October. She showed me her engagement ring. Matt had made it by weaving strands of heavy gold wire into a true-lover's knot (page 388 in *Ashley's*).

After talking with Lisa for half an hour, I turned to Matt and asked how such a smart and lovely woman could possibly hook up with a guy like him. I meant it as a joke. It's the sort of thing my friends and I say to each other all the time. But Matt hung his head, wounded, and said nothing. Later, after he'd had a few drinks, he approached me. Without rancor he said, "In answer to your question: I just got lucky. Real, real lucky."

I am a heel, I thought. *I am a shitheel.*

Hajo opened a case of homemade beer and passed bottles around. He'd brewed the beer in Traverse City and stowed a dozen cases of it below deck. It was in brown twenty-ounce bottles with labels he'd designed and illustrated himself. There was "Malabrew Deep Water Bilge Ale," "Bilge Reserve," and "9.5-Percent Cannon Fodder." All three were excellent.

Sitting regally in his deck chair with a beer in hand, Hajo obsessed about his cellphone. He'd been obsessing about it for three days and nights. He'd ordered it from a catalogue weeks earlier, and when it had not arrived by the morning of our departure from Traverse City, he gave serious consideration to delaying the trip. "I need my phone, man," he said. He said it many times. We learned later that it had arrived via UPS fifteen minutes after we left the dock in Traverse City. It was forwarded to Detroit, but was returned to Traverse City by mistake. Again it was sent south, this time to be delivered by hand by someone sailing a catamaran to Cleveland and back. We actually encountered that catamaran on the St. Clair River. It was motoring upstream while we went down. Hajo hailed the owner across the water and the owner shouted back that he hadn't found us in Detroit so he shipped the phone back to Traverse City.

Now a jaunty young man strode onto the dock, dressed in UPS brown, a package under his arm. Hajo leaped to his feet. "My phone!" he shouted.

A calm moment on Lake Michigan, near Sleeping Bear Dunes

The *Malabar*, moored in Traverse City, Michigan

Hajo at the helm, with Matt and Tim

The deck of the *Malabar*, heading north up Lake Michigan

Matt and Hajo in the main cabin

Voyageurs' canoe, on the north shore of Superior

Making camp on the north shore

Typical cobbled beach on the north shore

Lake Superior at Whitefish Point, looking north toward
where the *Edmund Fitzgerald* sank

The Whitefish Point Lighthouse

Hajo on the radio, Lake Huron

Meeting an ore carrier on the St. Clair River. It was a vessel much like this one that drove us aground that night.

Detroit skyline from the Detroit River

Matt and Lisa
on Lake Erie

Dave Mull with a nice
Lake Erie walleye

A round goby, one of many alien species that have invaded the Great Lakes

Dave Stone walking the shore of Long Point in search of artifacts

Pacific salmon like this
revitalized the Great Lakes
fishery in the late 1960s.

In a lock in the Welland

Stepping the masts in Oswego

In the Erie Canal

Removing storm debris from the canal

Top: Docked along the Hudson River in Albany

Bottom: The Pallisades of the Hudson River

Top: The George Washington Bridge, with the Little Red Lighthouse beneath it

Bottom: Tall Ships at the Manhattan piers

Hajo calling for a weather report on Long Island Sound

In Bar Harbor

The *Malabar*, dry-docked for inspection

Left to right: Matt, Luke, Jerry, Hajo, and Stan

He ripped open the package and withdrew a sleek, compact phone. It was state of the art. He leafed through the operating manual and within a few minutes had programmed the phone's ringer to play the beginning of Beethoven's Fifth Symphony. He called several people he knew and told them to call him back. Every time the phone rang, he held it in the air and insisted we listen to Beethoven. He kept missing his calls because he wanted to hear more of the symphony.

After dinner we sat on deck with Lisa and her father and talked. Walt was recently retired from his job as a chemist and was a gentle, patient, and thoughtful man, with a tendency to listen more than he talked. No problem. We talked enough for everyone. Hajo invited him to spend the night on board and he accepted happily.

In the morning, I went for a walk along the riverfront. The Detroit River and its banks have changed greatly since European settlement. After the *Griffin* sailed through in 1679, Hennepin wrote: "Those who will one day have the happiness to possess this fertile and pleasant strait, will be very much obliged to those who have shown them the way." It was Edenesque then, with meadows and woods ripe with wild grapes, nuts, and berries, and inhabited by so many bear, deer, elk, and bison that a brief hunting excursion got La Salle and his men all the meat they needed for their journey. Now, of course, it is city, the unsound heart of Detroit, packed with homes and businesses to the north of downtown and mostly industry to the south.

The river itself has been cleaned up dramatically since the 1950s and '60s, when it flowed thick with industrial and municipal wastes. It's become a vibrant fishery again, supporting most of the species found in the lakes. On slabs of concrete along the city's waterfront, I hunched down and talked fishing with some of the retired African-American men you can find sitting there almost any day. They prop their rods on forked sticks and sit in lawn chairs within reach of coolers filled with beer and sandwiches. Some of the best fishing is found virtually in the shadow of the Renaissance Center, a towering hotel and office complex built in 1977 by Henry Ford II as the great hope for the rejuvenation of the city; it now serves as world headquarters for General Motors. Nearby are the park at Hart Plaza, the hockey and baseball stadiums, the busy restaurants of Greektown, and a casino district. Not far away too are bombed-out crackhouses and abandoned ware-houses without a single intact window.

Pleasure boats roared past, staying close to shore to avoid the commercial traffic at the center of the river. While we talked, one old-timer caught a walleye and another a sheepshead. They said part of the appeal was you never knew what you were going to catch. They routinely took perch, pike, smallmouth bass, white bass, rock bass, sheepshead, carp, suckers, and wall-eye. Like their Canadian neighbors across the river, they called walleye "pickerel," and like anglers everywhere valued them as fine table fare. In the fall and spring they caught king salmon, coho salmon, pink salmon, brown trout, steelhead. Now and then they might hook a big muskie, but they usually couldn't land it. I asked how they felt about the quality of the water in the river.

"Lot better than it used to be," one of the men said.

"Shee-it," said another, "just look for yourself. See that bottom? All them rocks? Been a long time since anybody's seen bottom here."

Chapter 10

DETROIT RIVER/LAKE ERIE

Downriver Detroit ◆ On to Erie ◆ A Visit to Walleye World ◆
Invasion of the Zebra Mussels and Other Environmental Matters ◆
A Little Storm on Skeerie-Erie

The Detroit River downstream from Detroit to Lake Erie is a wasteland of
industries, some failed and failing, others apparently thriving. We passed
the mouth of the sluggish and dark Rouge River. Not far upstream Henry
Ford built the River Rouge Industrial Complex in 1919, where he mass-
produced Model-T's by the millions. He made practical use of the river,
building berths for steamships and discharge pipes for his factory wastes.
Other factories followed his lead, and for the rest of the century the Rouge
flowed alternately black with oil and bright orange with the discharge from
pickle plants. Public outcry in the late 1960s and early 1970s put a halt to
the worst of the point-source pollution and improved the appearance of the
river, but runoff from agricultural fields, inadequate municipal sewage treatment, and other non-point sources continued to contaminate the water.

Now a major cleanup effort is under way. By starting at the top of the
watershed and protecting the native wetlands that serve as a buffer and
regenerator, the entire river and the neighborhoods around it should benefit.
One of the significant environmental problems throughout the Detroit metropolitan area is stormwater capacity. Hundreds of square miles of land that
once served as a natural sponge to absorb and filter water are covered with
concrete and asphalt. They produce a deluge of runoff every time it rains,
flooding storm sewers and sewage-treatment plants, resulting in overflows
of raw sewage into the waterways. With millions of dollars already spent

and millions more projected, a state-of-the-art sewage and storm system is expected to soften the effects of runoff and prevent quite so much waste from entering the Rouge and thus the Detroit River and Lakes Erie and Ontario.

Downriver from Detroit much of the shoreline of what was once industrial wasteland and degraded wetlands has been the subject of recent conservation efforts. Eighteen miles of the U.S. and Canada shores, from Zug Island to Lake Erie, will be restored and protected as an international wildlife refuge. Parks and recreation areas might someday replace industrial sites.

We passed through the Detroit River's final "avenue" channel—a canalized shipping lane straight as a draftsman's rule and lined with cement riprap—and were met by the cool, unmistakable scent of big water. Suddenly we were at the mouth of the river and all of Lake Erie lay before us. The clear water of the Detroit River merged with the nearly clear water of Erie until they were indistinguishable. Ahead as far as we could see was milky-green lake bounded by pink horizons.

The day was sunny, shirtsleeve warm, freshened by a fine south breeze. We raised the jib, foresail, and staysail and soon were making 6.3 knots. The morning forecast had promised afternoon thunderstorms, with winds of twenty to thirty knots, waves three to six feet. It seemed unlikely now. But an update crackled over the radio from the Coast Guard: severe thunderstorm warnings and small-craft advisories for Lake St. Clair, the Detroit River, and western Lake Erie. We steered due east, but kept our eyes to the north and west.

Like a lot of people who have never seen Lake Erie, Hajo assumed it would be a cesspool. He was surprised to discover it is blue and clear and smells of clean water and beach sand, with a hint of approaching rain. It smells healthy.

Thirty years ago, of course, it stank. From the late 1950s until the early 1970s, it gave off a stench of human and industrial garbage that could be smelled many miles inland. Not much lived in the lake then except trash fish, algae, bacteria, and sludge worms. The algae, drifting in rafts two feet thick and miles across, was fed by a constant flow of agricultural, industrial, and municipal waste being dumped into the lake. A little algae is a good thing, but too much sucks the oxygen from the water. Outbreaks of a toxic blue-green algae, *Microcystis*, proved deadly to fish and aquatic plants and caused hepatitislike symptoms in humans.

Anyone who knew Erie in those days can tell horror stories. One guy I met remembers when he was a young man motoring from Maumee Bay to West Sister Island, traveling the entire twenty miles through a mat of bright green algae so thick the trail his boat made was still there when he returned at the end of the day. Back then, he says, nine out of ten fish you caught were sheepshead, a bottom dweller generally considered a trash fish. Beaches all around the lake were fouled with mounds of detergent suds as high as your head. The place smelled like a sewer.

But now Erie's unpleasant olfactory past seemed remote. I had recently spent two weeks driving around the lake, meeting biologists and anglers, getting on the lake whenever the opportunity arose. One day I went out in the deep water around the Bass Islands near Port Clinton—and in Erie's western basin, deep water means thirty or forty feet—where anglers from all over the world gather to fish for trophy-size walleye. The water this far out was as clear as Superior's, with rocks visible on the bottom thirty feet down. Closer to shore, three days of storm had churned up silt and clay and created a mocha-colored band a mile wide from Toledo to Cleveland. Fishing in the obscured water would have been unproductive. Out here it was a whole lot better.

I was a guest on a boat skippered by Jim Stedke, a fifty-one-year-old tournament angler who's been fishing the lake most of his life. With us was his friend Scott Stecher, a thirty-seven-year-old former high school biology teacher who had recently quit teaching to design and manufacture fishing lures. Rounding out our group were two old friends of mine—Ron Barger, owner of a marketing communications firm in Niles, Michigan, and Dave Mull, editor of *Great Lakes Angler* magazine.

Recreational walleye anglers on Erie divide into two camps, trollers and drifters. Trollers use multiple rods, outriggers, planer boards, downriggers, sonar—applying all the technology available to their search for big, open-water walleye. Drifters are traditionalists, fishing the time-honored way on Erie: locating submerged reefs of gravel and limestone and drifting with the wind over them, using spinning rods to cast jigs, spoons, and in-line spinners. Drifters often catch the most fish—whole boatloads of three-pounders—but trollers catch most of the biggest ones.

At first we had trouble finding them. Jim took us to an area that had produced well a couple days earlier. The sonar screen was blank except for occasional Christmas trees of bait fish. We pulled in the lines and motored to the channel between Middle Bass and South Bass islands. Again, schools of walleye were here a few days ago. But not now.

"This is like the bad old days," Jim said. "I started fishing the lake in the 1950s, with my dad, and got my own boat in 1963. Back then we had to fish hard to find any walleye at all. I remember one day we caught thirteen and thought we'd gone to heaven. That was unheard of in those days."

By the late 1980s and early 1990s, walleye fishing in Erie was arguably the best in the world. The lake held so many adult walleye that the state of Ohio raised the daily bag limit from six to ten per angler. Commercial fishermen, especially in the Ontario waters of the northern half of the lake, netted them by the ton. If you didn't limit out, it was considered a bad day. Even now, after several years of declining catches (and with the daily limit in Ohio waters reduced to five), nobody expected to get skunked. Which made our present lack of success remarkable.

Serious anglers love this type of challenge. Jim and Scott, eyes gleaming, engaged in a running analysis of water conditions, weather, the movement of the schools. Lake Erie walleye do not necessarily seek bottom structure, as they do in smaller lakes. Commercial fishermen discovered decades ago that the fish could be caught in nets set just beneath the surface in the open lake. Sport fishermen followed suit, going after them with downriggers and planer boards. With the clearing of the water starting in the 1970s, and especially after the infestation of zebra mussels, which filter the water with stunning efficiency, the open-water fishery is increasing in importance. Fewer walleye are found on shallow reefs now and more are suspended in schools over deep water.

Finally, over the radio, came a tip. A charter captain notified Jim that he was catching walleye from the west side of Kelley's Island. Bait fish were stacked up on a bank there in thirty-five feet of water, and walleye were feeding on them. Soon we fell into line with two dozen other boats trolling the bank. Our lines were in the water less than five minutes when the first rod released and a five-pound walleye came to the net. In two hours we caught fifteen, the biggest about eight pounds, most about five. I'd never seen so many big walleye caught in such a short time. Jim apologized for not leading us to bigger ones—the water is famous for ten- and twelve-pounders. That northeast wind, he said.

For more than a decade the quality of the fishing in Erie has been astonishing, among the best anywhere for walleye, bass, muskie, yellow and white perch. For generations in the nineteenth and twentieth centuries the game-fish of choice were yellow perch, walleye, and blue pike, a smaller cousin of the walleye now considered extinct. Commercial fishermen filled the holds of their trawlers every time they went out on the lake. Filleted and flash-frozen, the fish were shipped by train to cities across most of the eastern half of the continent. Small boats crowded every shore, entire families taking to the water in the evenings to cast nightcrawlers and minnows, not considering the trip a success until they had filled every available tub and cooler. Blue pike especially were ridiculously easy to catch and so abundant that it was inconceivable they would ever disappear.

Then, late in the 1950s, the fishery collapsed. Nobody heeded the warnings: declining commercial catches, mysterious blooms of algae, an ecosystem thrown into disorder by the introduction of exotic species like the sea lamprey, alewife, carp, and rainbow smelt. The collapse didn't occur overnight, but it happened quickly enough to leave people shocked and angry. Catches dropped precipitously until, by 1960, walleye were scarce and the blue pike had vanished. The lake, once pristine, could hardly be recognized.

Its condition caught most people by surprise. Industry had always dumped wastes and cities had always dumped sewage into Erie because it was easy and cheap to do so and because the public naively believed that such a large body of water—9,900 square miles, larger than the state of Vermont—could absorb garbage faster than it could be dumped. Besides, the lake's water was being continuously flushed out the Niagara River, to Lake Ontario, and out the St. Lawrence to the Atlantic, the ultimate receptacle. Sure, Erie suffered some pollution near rivermouths and cities. But polluted beyond salvation? It was unimaginable—until the day it became true.

Suddenly the blooms of algae that had come and gone came and stayed. The giant *Hexagenia* mayflies that for as long as anyone could remember had hatched in such numbers that snowplows were sometimes needed to clear them from roads along the lake, stopped hatching altogether. Beaches filled with stinking windrows of dead fish, heaps of decomposing algae, small mountains of detergent suds. Counts of fecal coliform bacteria went off the charts. Limnologists studying oxygen levels discovered that 2,600 square miles of bottom water in the central part of the lake contained no dissolved oxygen at all. And nothing lived there.

The evidence could no longer be ignored. It stank too much to ignore. The lake was going to hell. And not just parts of it. The entire lake, from surface to bottom, was undergoing a massive ecological collapse.

Biologists diagnosed the problem as "eutrophication"—a too-nourished state caused by all those nutrients, especially phosphorous, washing into the water. Lake Erie has always been more eutrophic than the other Great Lakes—it is warmer and shallower, surrounded by land layered with rich loam and clay that naturally finds its way into the lake—but this was extreme. Some limnologists calculated that the runoff from industrial processes, from farmers' fields, and from municipal sewage plants (many of them consisting only of primitive settling ponds designed to keep the most obvious raw solids out of the lake) had added the equivalent of fifty thousand years of age to Erie. The popular press picked up the story, and suddenly headlines everywhere declared the lake was dying. It became the American Dead Sea. Cleveland, the industrial behemoth responsible for significant amounts of pollution, was "The Mistake by the Lake." By 1969, when gasoline on the surface of Cleveland's Cuyahoga River ignited and burned, "Save Lake Erie" had become the rallying cry for a burgeoning environmental movement.

Erie's return to health is now recognized as one of the greatest environmental victories in North America. The lake was saved by public outcry that led to legislation, most notably a trio of acts signed into law in 1972.

The U.S. Clean Water Act, the Canada Water Act, and the Great Lakes Water Quality Agreement established standards for water quality, mandated the cleanup and protection of the Great Lakes, and set penalties against industries that polluted.

The effect on Lake Erie was immediate and dramatic. The lake averages only sixty-two feet deep and is flushed clean every 2.6 years, fastest of any of the Great Lakes (Superior, by contrast, requires 191 years). So, although the ecological damage had been great, the first stages of recovery were rapid. By 1975 the algae blooms had ended, dissolved oxygen had returned to bottom waters, and public beaches were again safe for swimming. The once soupy green water started to become clear again. Blue pike were gone forever, but perch and walleye returned.

Nobody seems happier about that than Carl Baker, a cheerful, easygoing man in his sixties who until 1994 was supervisor of Lake Erie Fisheries Research for the Ohio Department of Natural Resources. Though officially retired, he still does field research and spends as much time as he can on the lake. While we fished together one day on Maumee Bay, at the west end of Erie, he told me about his first look at the lake, in the early 1960s.

"I was a young biologist, fresh out of college and eager to save the world," he said. "Then I saw Lake Erie. It looked like it was covered with spilled paint—giant swirls of green and yellow and red algae everywhere, as far as I could see. Nobody fished in it, there was no point. I almost gave up and left."

Instead, he devoted his career to helping restore the fishery. Though he waves off the statement, some people claim he's the man most responsible for the walleye boom.

I asked Jim Fofrich, a charter captain out of Toledo who for three decades has been active in conservation efforts on Erie, what role Carl Baker played in the health of the fishery. "Here's one example," Jim said. "When the lake cleaned up a little and the fish started coming back, commercial gill netters went to work with a vengeance. They weren't allowed to net walleyes, so they went after chubs and white bass and smelt and anything else the market would bear. Trouble was, they netted thousands of walleyes in the process, and because they weren't allowed to keep them, they just threw the carcasses back into the lake. Carl did a study on non-targeted fish mortality, proved that tons of walleyes were being killed senselessly, and was able to get gill nets declared illegal in Ohio waters. It's what made it possible for the walleyes to come back in such big numbers."

That day I had joined Carl and a party of several anglers on Jim Fofrich's boat in Maumee Bay, at Erie's extreme western end. Fisheries biologists get excited when you mention Maumee Bay. It and the Maumee River are often described as the primary walleye "factories" in Lake Erie—this in spite of their reputation as being among the most polluted portions of the Erie watershed. In the 1950s and '60s, the Maumee River, which contributes about three percent of the water volume of the lake, contributed nearly half the sediment load being dumped into it. Recent research suggests that walleye reproduction is occurring on reefs in the bay, places that were assumed unsuitable because they're covered with sand and silt. But every spring in recent years the bottom in certain parts of the bay has become loaded with walleye spawn. Since walleye need a firm substrate on which to deposit their eggs, it's been suggested that the spawning might be occurring on beds of zebra mussels. Carl Baker and other researchers want to know if that's true. That day on Fofrich's boat, Carl was clipping fins on walleye for DNA analysis to see if distinct populations of fish were returning to the same reefs year after year.

The reef we fished is a hump rising from fifteen feet of water and was formed by debris dredged from the shipping channels. It was about the size of a city block. We motored upwind to the front edge of it, killed the engine, then drifted across, casting as we went. Eight of us fished; Jim mostly handled the landing net. It took about fifteen minutes to pass over the hump, all of us casting furiously, and we boated a dozen fish. Other boats fished the same structure in the same way. Everybody caught fish.

Maumee Bay is ringed with an industrial skyline dominated by the cooling towers of a pair of nuclear power plants—massive Fermi south of Detroit and Davis-Besse east of Toledo, each with a cloud of steam flagging perpetually from its top. Factories, derricks, loading docks, and freighters line the shore and stagger the horizon, like the industrial wasteland in *Blade Runner.*

Yet the fishing is spectacular. Life flourishes, even in the shadows of the industry that once nearly killed the lake. In this single afternoon I caught more walleye than I'd caught in years. And, except for a few runts and the twenty mature males we measured, fin-clipped, and released as part of Carl Baker's study, we kept every one of them. Filled our coolers, in fact.

Baker warned me against being too optimistic about the health of the lake. Those of us who have vivid memories of the days when "Erie" was synonymous with environmental disaster, when it was widely believed that anyone who fell into that cesspool of poisons would have his flesh stripped from his bones by chemical piranha, are tempted to view the lake's comeback as miraculous and permanent. But Erie has serious problems yet.

The most visible threat is biological. Exotic species have been invading the Great Lakes since the nineteenth century. In 1869, purple loosestrife, a plant native to Europe and Asia and introduced as a decorative plant to eastern North America, appeared in wetlands around the Great Lakes and began crowding out cattails and other native plants. In 1873 came the alewife, a small forage fish that would have a big impact in the next century. In 1879, the common carp was stocked in the Great Lakes as a potential gamefish and promptly began rooting up aquatic plants and muddying shallows.

The invasions multiplied after improvements were made to the Welland Canal in 1919 and with the opening of the St. Lawrence Seaway in 1959. Suddenly alien species were launching a three-pronged attack. Some, like rainbow and brown trout and Pacific salmon, were introduced intentionally. Others, like the sea lamprey and alewife, swam upstream from the Atlantic via the St. Lawrence or Erie Canal. Still others, like the zebra mussel and the European ruffe, stowed away in ballast tanks of oceangoing ships arriving from Europe and Asia.

Today the Great Lakes are home to more than 140 species of animals and plants that have no business being there. When non-native species are introduced to a new habitat they often devastate native species, alter and degrade the habitat, and otherwise create enough damage to deserve their reputation as biological pollutants. Some invaders have found the Great Lakes a nearly ideal place to live. Without predators, parasites, pathogens, and competitors, their numbers have exploded.

Many of the recent invaders thrive in the warm, fertile waters of Erie. The round goby, a small bottom-dwelling fish native to the Caspian and Black seas region, is undergoing a population explosion and is now found by the thousands on rocky reefs once inhabited by walleye. The Eurasian ruffe, a spine-backed, perchlike fish from Eurasia, competes with other species but is rarely preyed on itself. The spiny water flea, a tiny crustacean native to northern Europe, is competing with native crustaceans and appears to be winning, largely because it repels predators with its long, barbed tail. The rusty crayfish, native to streams in Ohio, Kentucky, and Tennessee and introduced into the Great Lakes by anglers using them as bait, are outcompeting indigenous crayfish. Dozens of other animals (and many plants) have entered the Great Lakes and are poised for infestation.

Of recent invaders, the zebra mussel has made the greatest impact. This pistachio-size mollusk, native to the Caspian Sea, has been joined more recently by a close relative, the quagga mussel, which lives in slightly deeper water. Both mussels were probably brought across the Atlantic in the ballast tanks of ships. First discovered in Lake St. Clair in 1988, the zebra mussel is now ubiquitous, clinging to submerged rocks, logs, and man-made structures throughout the Great Lakes, as well as in its connecting rivers (including, via canal, the Mississippi). It's appearing also in many inland lakes within the watershed. So far Lake Superior has been protected by its cold water and has suffered only small infestations.

Lake Erie has been hit hard. Billions of zebra and quagga mussels carpet the bottom in masses as large as townships. The mussels are prolific—each female produces up to a million eggs a year. They feed by filtering plankton and other organisms from the water, each mussel straining all the suspended matter from a liter of water a day.

When the mussels first appeared, there seemed little reason for alarm. In fact, the mussel's extraordinary ability to clear the water was seen as a benefit to tourism, especially on Lake Erie. Waters that had been turbid were suddenly crystalline, partly due to the general cleanup of the lake in the past couple decades, partly because of the filtering of the mussels. Scientists are amazed at the recent "ultra-oligotrophication" of Lake Erie—its sudden change from a highly productive to a less productive state. But the clear water has come at a price. Intake pipes at power plants and water-treatment facilities must be cleaned regularly, at high cost, and all those sharp mussel shells on the lake bottom require bathers in many shoreline areas to wear protective footwear. Long-term effects may be even more costly. After a ten-year absence, recent outbreaks of the toxic algae *Microcystis* are possibly being caused by zebra mussels, which recycle phosphorous and perhaps make it more available to the algae. Even more disturbing, outbreaks of botulism in 2001 and 2002, which killed thousands of loons and other waterfowl in Lake Erie, have been traced to zebra mussels and the round goby, which feeds on them. Biologists who two years ago were reluctant to speculate whether zebra mussels affect other organisms are now suggesting that they could undermine food webs throughout the Great Lakes.

The issue is complex. Zebra mussels feed heavily on phytoplankton, the foundation of the "biomass pyramid," and are almost certainly outcompeting other plankton eaters. It requires a thousand pounds of phytoplankton to grow a hundred pounds of zooplankton, a hundred pounds of zooplankton to produce ten pounds of forage fish, and ten pounds of forage fish to support a single pound of walleye or other large predator. Take away the plankton and you take away the walleye. But in this case, will walleye and other top-of-the-chain predators be affected? Zebra mussels filter virtually all organic particles from the water, everything from clay to algae, but much of it ends up encased in tiny mucus-covered pellets called pseudofeces, which are expelled to the lake's bottom. Gradually the pellets decompose, are fed upon by bacteria and perhaps other organisms, and again enter the ecosystem. Some zooplankton and other small animals are suffering from the change, but others could be benefitting. At least one species of scud—a small crustacean sometimes called a freshwater shrimp, and an important prey species for fish—seems to be thriving on the bounty of pseudofeces in Lake Erie.

But the recent disappearance of another shrimplike crustacean, a half-inch-long amphipod called *Diporeia,* has caused concern and bafflement among biologists. At least as late as the 1980s, amphipods were found in the bottom mud of all the Great Lakes in numbers as high as ten thousand to every square meter. In the early 1990s, however, researchers were startled to notice the population declining. Samples taken from a five-mile-wide strip of lake bottom along the southeastern shore of Lake Michigan, from Chicago to St. Joseph, indicated a sixty to ninety percent decrease in the amphipod population. By the end of the decade, the researchers could find none. Not even one. In waters off Manistique, Michigan, at the northern tip of Lake Michigan, bottom that contained thousands of amphipods in 1997 had none in 2000. Whitefish, which feed heavily on the amphipods, were discovered to be feeding instead on zebra mussels—or trying to feed on them. Their stomachs were packed with mussel shells but the whitefish were so malnourished that commercial fishermen couldn't get enough fillets off them to sell.

Meanwhile, *Diporeia* had disappeared from Lake Erie. None of the amphipods could be found anywhere in the lake. In bottom sediment where they had normally accounted for seventy percent of the living biomass, they now accounted for zero percent.

While it's too soon to call *Diporeia* a canary in the mine, something is clearly wrong. In a statement to the press in 1997, Tom Nalepa, a biologist for the National Oceanic and Atmospheric Administration (NOAA)'s Great Lakes Environmental Research Laboratory in Ann Arbor, placed the blame on zebra mussels. He was among the first to speculate publicly that the mussels might be extracting a nutrient essential to *Diporeia,* or that they were somehow releasing a pathogen lethal to the amphipods, or that the mucous in their pseudofeces was toxic. Zebra mussels filter diatoms from the water, and diatoms, when they settle to the bottom of the lake, are the primary food source for amphipods. Nobody could explain why the amphipods were disappearing from portions of Lake Michigan where the zebra mussel is still scarce, but nonetheless there seemed ample evidence of a cause-and-effect relationship.

"What's happening," Nalepa said, "is energy that used to support amphipod growth is now being turned into zebra mussel tissue. Many species of fish, and particularly young fish, readily eat amphipods, but few species can use zebra mussels for food. There's concern that such a short circuit in the food chain could lead to declines in a number of fish, including perch, alewives, sculpin, bloater, and smelt, with possible secondary effects on trout and salmon."

Already Lake Erie is incapable of supporting the numbers of walleye that lived there in the 1980s. Most forage fish seem to be finding alternatives to *Diporeia*, and for now at least the fishing remains quite good. But anglers are being warned to expect gradually reduced catches in coming years. Fisheries researchers estimated that the total number of walleye in the lake fell from a high of a hundred million in the early 1990s to twenty-five million by 2000. Under pressure from U.S. sportfishing groups, Ontario agreed in early 2001 to slash its commercial fishing quota of walleye by ninety-four percent for at least three years. Since Ontario commercial fishermen previously took seventy percent of the walleye caught from the lake every year, it is hoped that this reduction of quotas will allow the population to rebound.

Serious as they are, biological pollution and reduced productivity are not the only problems facing Lake Erie and her sister lakes. In 1989, when commercial and recreational fisheries—and the local economies that depend on them—were finally thriving again, the National Wildlife Federation created enormous controversy by announcing that eating those fish was hazardous to your health. How could this be? The lakes were cleaner than ever. Most of the dumping of wastes had been halted. How could the fish be poisoned?

Blame Pollution Abatement Services, an Oswego, New York, disposal company that in the 1960s and 1970s dumped and buried tens of thousands of barrels of toxic chemicals on sites adjacent to Wine Creek, a tributary of Lake Ontario. When Pollution Abatement Services went bankrupt in 1977 and abandoned the site, they left their mess behind: a leaking, stinking brew of dioxins, PCBs, DDT, Mirex, and other chemicals demonstrated to be carcinogenic and bioaccumulative, all of them leaching into Wine Creek and thus into Lake Ontario.

Blame Dow Chemical of Midland, Michigan, a company that until the mid-1980s routinely discharged about forty contaminants, including a form of dioxin ten thousand times more potent than DDT, into the Tittabawassee River, which flows into the Saginaw River, a tributary of Lake Huron's Saginaw Bay. The Tittabawassee is still so polluted with PCBs and dioxins that anglers are warned to avoid eating any carp, catfish, or white bass that live in it.

Blame the Hooker Chemical Company for dumping twenty thousand barrels of dioxins, Mirex, and other toxins in a site adjacent to Lake Michigan near Montague, Michigan, and for burying 21,800 tons of toxic materials in Love Canal near Niagara Falls.

Blame Western Reserve Mining Company for dumping so much asbestos-contaminated taconite slag into Lake Superior from 1955 to 1980 that the entire western end of the lake is carpeted with carcinogenic asbestos. Gradually the layer of asbestos has been entombed beneath sediments, but whenever the bottom gets stirred, clouds of fibers rise again.

Blame paper mills and water-treatment plants that until 1982 poured industrial sludge and raw sewage into Green Bay. Blame the steel and chemical plants in Gary, Indiana, for dumping everything from cyanide to motor oil to heavy metals into the Grand Calumet River and southern Lake Michigan.

For more than a century uncountable tons of industrial chemicals, byproducts, and heavy metals were washed or blatantly dumped into the Great Lakes. The water quality acts and agreements and the banning of DDT and PCBs in the 1970s were big steps in the right direction, but the problem with heavy metals and many other toxic pollutants is that once they've entered an aquatic ecosystem, they remain there for decades and sometimes centuries. Researchers discovered that although the water itself in the Great Lakes now contains only traces of the chemicals and metals, the bottom sediments in many places are poisoned with dangerous quantities. Sometimes unbelievable quantities. When sediment was removed from Lake Michigan's Waukegan Harbor, north of Chicago, it was found to contain 500,000 parts per million of PCBs—a fifty percent concentration.

One day I took a ride in a small boat to the mouth of Lake Erie's Black

River in Lorain, Ohio. It was an ugly day. The barometer was falling and the wind was hard from the northwest, shoving big rollers into the harbor from the open lake. Clouds scudded past, low and gray, the precise color of pessimism. The Black is deeply brown and lined with heaps of coal and rusting machinery and smoking industrial wreckage of the sort that conjures the environmental nightmares of half a century ago. It is one of Erie's most contaminated tributaries. The sediment at the bottom of the harbor and the lower river is loaded with enough poisons to fill a spice rack from Hell: PCBs, dioxins, furans, organochlorine pesticides, polycyclic aromatic hydrocarbons, and heavy metals such as cadmium, lead, and mercury. The water never has a chance to be flushed clean. Every time the bottom is stirred by dredging, boat traffic, currents, storms, and bottom-dwelling organisms, a new slurry of horrors is flushed free.

Once those horrors are stirred up, they follow an insidious route through the food web. Toxic substances accumulate in the soft, fatty tissues of animals. As the contaminants work their way up the food chain, they biomagnify, becoming more concentrated in predators than in their prey. A zooplankton might contain only a minute amount of dioxins in its body, but the emerald shiner that feeds on thousands of zooplankton in its lifetime accumulates a greater concentration of the chemicals. The salmon that eats the shiner accumulates more yet, and the bald eagle that eats the salmon accumulates the most of all—a concentration more than a million times greater than is found in the water itself.

What is true for the eagle is true for humans. People who regularly eat fish from the Great Lakes run the risk of accumulating potentially harmful contaminants. It is uncertain what effect many of the substances have on human health, but most scientists agree that they increase the risk of cancer and birth defects, especially for children and fetuses.

In the 1980s and '90s, after much had been done to stop the flow of contaminants into the lakes, researchers noticed a disturbing trend. Contamination levels had dropped in the years immediately following the chemical bans, but the concentrations in wildlife and humans had leveled off, and in some cases were climbing again. About the same time, researchers studying isolated populations of fish in an inland lake on Lake Superior's

Isle Royale, far from any known source of water pollution, found the fish contaminated with PCBs and the pesticide Toxaphene. They concluded that those contaminants could have arrived from only one source: the sky. Discharges from factory stacks and house chimneys, from smoldering land-fills and automobile exhausts, from coal-fueled power plants and gas-powered factories—some in the United States and Canada, others from as far away as Russia—were being carried around the world by high-altitude winds and carried to the ground by rain and snow. More recent studies have suggested that not only are the Great Lakes subject to atmospheric depo-sition, they're creating it. Sediments in southern Lake Michigan are so polluted with PCBs that the chemicals rise through the water and escape into the air. Brian Eadie, senior scientist at the Great Lakes Environmental Research Laboratory in Ann Arbor, describes Lake Michigan as a chemical reservoir busily "exhaling PCBs" that are eventually deposited on land and water elsewhere.

Today, a careful balance has been struck between public safety and local economies. Fisheries departments certainly don't want to frighten anglers away—sportfishing is big business in the Great Lakes—but neither do they want them to get sick. The compromise has been to issue carefully worded consumption advisories. The fishing is great, they say, and eating fish has been shown to reduce heart disease and other illnesses, but wise anglers limit the amount they eat. In Ohio, for instance, it's suggested you consume no more than one meal a week of Lake Erie's walleye and perch, one meal a month of smallmouth bass. Somehow the committee-derived wording or the constant repetition have robbed the warnings of impact. Even the stern-est of them—that children and pregnant women should not eat certain fish because of the risk of birth defects—often go unheeded. Nobody knows how many of the millions of anglers who fish the Great Lakes take the warnings seriously. I've talked to dozens in every state and Ontario who admit frankly and even proudly that they ignore the advisories. Some anglers trim away the bellies and other fatty parts of the fish when they clean them, as is recommended, but the general attitude seems to be that the threat is exaggerated, that what you can't see can't hurt you, that there's nothing to worry about at all.

But people who have studied the contamination problem find plenty to worry about. The International Joint Commission, which was created by the United States and Canada to advise on Great Lakes water issues, has identified forty-two toxic hot spots called "Areas of Concern" (AOC) that contain highly contaminated sediments. But identifying an AOC is relatively easy; cleaning it is hard. First, it's expensive. The Environmental Protection Agency recently released a study concluding that the high cost of "remediating" Great Lakes sediments was more than offset by benefits such as preventing cancer in humans and reducing environmental impacts on tourism. Yet getting rid of polluted sediment from a single harbor or rivermouth can cost millions of dollars and take years to complete. Dredging is the only practical way to remove the sediments, but dredging often stirs up more problems than it solves. For every bucketload taken out, many more are set free as drifting clouds of silt. And the muck that is removed must be dumped somewhere. Nobody wants poisoned soil deposited in their neighborhood. The solution so far has been to bury the sediments in sealed landfills or cook them at temperatures high enough to destroy contaminants. But those methods can cost thousands of dollars for every cubic yard of sediment.

From the middle of Erie we could see Ohio's Bass Islands, low and vague with haze. On South Bass, the Perry Memorial rose above the horizon like a middle finger thrust at Canada. I wondered if the gesture was intentional. For how many years did resentment simmer after the War of 1812? These greenish-blue waters were the scene of one of the most dramatic naval battles in North America. In September 1813, the young naval officer Oliver Hazard Perry and his fleet of ten ships destroyed the British fleet there and helped win the war for the United States. After the battle he sent his famous message to his commander on shore: "We have met the enemy and they are ours."

Lisa took the helm and stayed there much of the way across the lake. Already she had proven herself a knowledgeable and capable sailor. And she was a patient teacher. She showed me how to belay properly and how to identify some of the most important lines and sheets.

To the north, reaching down from the Ontario shore, was Point Pelee National Park, considered one of the five best bird-watching destinations in North America. On May mornings it's not unusual for visitors to this "migrant trap" to see thousands of warblers, tanagers, orioles, flycatchers, hummingbirds, hawks—up to 357 species of birds have been counted. They crowd the Point, often sitting on the ground to rest after their exhausting night flight over the lake. In the fall they stage at Pelee, waiting for favorable winds to push them south. The protected, marshy Point is a magnet for many kinds of wildlife, though not as many as Charlevoix encountered in 1721, when he wrote: "There are a great number of bears in this country, and more than four hundred of these animals were killed last winter on Pointe Pelee alone."

Past Point Pelee the lake widens and land slips from sight. We sailed downwind then, enjoying a splendid day. But anviled stormclouds towered to the north. They were headed for New York, same as us.

That evening for dinner we dined on grilled quail breasts and sautéed fresh vegetables in a cream sauce over pasta. The night before we'd had beef brisket in marinara sauce. Tim was outdoing himself. In Detroit he had reconsidered his earlier misgivings. Once we were on Erie, he pulled me aside and apologized for wanting to quit the boat. "I'm glad I stayed," he said. "I have no desire to ever repeat this trip, once is enough, but I'm happy I came along. It's an exploration. I'll never forget it."

After dinner we lowered the mainsail and jib, leaving up only the foresail. I went to bed, determined to get some sleep before my midnight watch.

I was awakened in darkness by Harold standing at the open hatch shouting: "All hands on deck!" Fearsome words. I lurched out of my cabin in my longjohns. Matt and Tim burst out as well, their eyes round. I was aware of the boat heaving in several directions. High in the rigging was a sound like animals screaming.

Uptop: chaos. A squall had swept without warning across the lake and caught us broadside. Wind pounded the sail, making the canvas luff violently. The boat rushed along, heeling hard to starboard, and waves washed over the scuppers. Hajo stood at the helm, fighting to control it, shouting at us to lower the foresail. Now! Now! We lowered the sail, reefed it to the boom, and secured all the lines.

Then the squall passed. In a few seconds the wind dropped from twenty-five knots to ten and the chop flattened. Stars appeared through tatters in the clouds.

"What the *hell* was that!" Hajo shouted. "I can't believe it! That wind came out of nowhere! If we'd had all our sails up it would have knocked us down! Skeerie-Eerie, man!"

Erie is famous for its sudden storms. I met a man in Port Clinton who was out on the lake one day in his fishing boat when a wind came up so strong that it capsized him. He insisted that his boat was not turned over by waves—the wind arrived so fast there were none—but by the wind itself prying under the hull like a spatula and flipping it. He was trapped underneath for nearly an hour, his boat turtled over him, before somebody rescued him.

The sky cleared and the wind settled as if nothing had happened, though the lake maintained a two- to three-foot chop. We were shaken, but the danger was past, and finally Hajo, Harold, and Tim went below to sleep. Matt and I stayed up alone, Matt at the helm. I sat in the bow and watched the waves emerging from the darkness, their crests breaking and flashing white.

I was curious to see how Lake Erie's waves differed from those I knew so well on Lake Michigan. The size of waves always depends on a variety of factors, including the wind's strength, duration, and fetch, and the depth of the water. Oceanographers calculate that a wind blowing at twenty knots for twelve hours over a fetch of a hundred miles will generate waves up to six feet in height. But if the wind speed doubles to forty knots over the same duration and fetch, the waves will grow to thirty feet, a fivefold increase. This disproportionate gain is the result of a mathematical quirk: the force of the wind is directly proportional to the square of the wind speed. Waves grow exponentially.

They also grow in response to current. The height of waves can double when they meet a contrary current of only two or three knots. This is significant in places like the Mackinac Straits, and it can be deadly in ocean straits, where tidal currents form powerful, temporary rivers. Waves meeting current in such places often creates legendary turbulence.

Waves also increase in size and steepness when they pass over shoals, as they do often in the Great Lakes. Lake Erie is only 62 feet deep on average, and its deepest spot is only 210 feet. The lake divides neatly into thirds—the Western Basin, Central Basin, and Eastern Basin. Most of the deep water is in the east. The western two-thirds can be considered a single big shoal.

About three in the morning, Matt and I noticed a black wall of forest along the north shore. But that was impossible; we were too far away to see shore. A crack of sky appeared behind the wall, then grew wider. We realized with a sickening sense of foreboding that the wall we were watching was not forest, but clouds. Squall line approaching.

We stowed gear below deck and closed all the hatches. We wanted to put on our foul-weather clothes, but there was not enough time. I stood at the helm as the wind struck in a cold blast, the line of clouds rushing close behind. The first gust blasted us broadside and caught the bare masts with such force that the wood groaned and the boat heeled far to starboard. The rigging screamed. Dishes crashed below in the galley. Matt had to grab the rail to keep from falling.

The air around us filled with spray. Confused seas rushed from the darkness and smashed into us—heavy swells from the west, wind waves from the north, both slamming against our hull and washing over the scuppers.

Then it ended. Hajo ran on deck, pulling on his jacket and yelling for a damage update, but the squall had already passed to the south and the clouds above us were breaking up. The wind died, and the moon came out. It stayed out the rest of the night.

Skeerie-Erie, man.

Chapter 11

LAKE ERIE

Downwind on Erie ◆ The Beachcomber of Long Point ◆ The Welland
Canal and the Invasion of the Sea Lamprey ◆ Salmon Come to the Lakes

In the morning the sky above Lake Erie was crisp and blue, scrubbed clean
by the storm. A daisy-ring of distant land clouds circled the lake. "Like a
halo," Hajo observed. We raised the mainsail and foresail and beat downwind on seas three to five feet. The waves made the boat buck lethargically.

Early into the trip, Hajo and Matt had joked that Lakes Michigan and
Huron were too colorful. "It's scary—water's not supposed to be this blue,"
said Matt. Now they compared Erie's greenish-blue to Corpus Christi
Harbor, the Gulf of Mexico, Islamarada. They thought the short swells
were a little like Chesapeake Bay's, but not quite. They had never seen
waves like them.

Erie is a bad place to be in a storm. Its southwest-to-northeast orientation
allows the prevailing winds to funnel its 241-mile length, a fetch sufficient
to build dangerous seas. The shallow water drags at the bottoms of waves,
causing the crests to overrun themselves and collapse, turning the surface
into a vast breaker zone. Unforecasted storms appear with terrifying quick-
ness; sudden winds have reached a hundred miles an hour, stripped boats
of their masts, and capsized them. Few safe anchorages can be found, so a
vessel, once committed, must keep going until it runs out of wind or runs
out of lake.

That's true for the other four lakes as well, which no doubt explains why
so few pleasure craft are seen in the open lakes. Near harbors we encoun-

tered so many sailboats, powerboats, and fishing craft, that I was not surprised when I later learned that the eight Great Lakes states are home to one-third of all the registered recreational boats in the United States. Thousands of them are concentrated in the harbors in Chicago, in Grand Traverse Bays, and in Green Bay, Saginaw Bay, Port Huron, Detroit, Cleveland, Toronto, and other protected waters near population centers. But get away from sheltered water and you see few boats.

The most prominent natural feature along Erie's north shore is Long Point, a twenty-mile-long sandspit protruding into the lake from Ontario. Much of the interior of the point is forested, but the protected inside shore is an everglades of marshes and swampy islands. The western, windswept shore is lined with sand dunes and beaches deposited there by thousands of years of wind and current. Most of it is a UNESCO Biosphere Reserve, set aside for research and conservation, and not accessible to the general public.

Sailors are wise to be cautious around Long Point. It is a snare reaching to the center of the lake, with shoals of sand surrounding it on every side. If a vessel runs aground there in heavy seas, it is usually lost.

La Salle and the *Griffin* nearly ended their maiden voyage at Long Point. They had just left the Niagara River and were heading west up the lake when they encountered fog so dense the crew could see nothing around them. They knew by their compass heading that they should be on a safe course, but through the fog they could hear the muffled crash of surf. They took soundings with a lead weight. The bottom was rising. Land was ahead. They lowered their sails and drifted to a halt and waited. When the fog lifted, they saw Long Point dead ahead and were able to raise their sails and steer around it. In the years to come, hundreds of ships would be less fortunate.

Mention shipwrecks to anyone who lives near Long Point and they'll bring up Dave Stone's name. "The Beachcomber," they call him. He's been exploring the Point since 1933, when he was a kid staying at the family cottage down the shore, and has lived most of the summers of his adult life in a beach cottage at its base. He's a scuba diver who has dived on hundreds of wrecks in Lake Erie. He's also an amateur historian and archaeologist with a special permit to explore the Biosphere's shores, and is the author of two books about the area. Nobody knows Long Point better.

At his invitation, I had visited him one morning the previous summer. I knocked at the door of his cottage and got no answer, so I walked around to the front. On the beach was a man dragging an aluminum boat through the sand. I knew he was seventy-some years old, but he didn't seem it. He was small and wirey, barefoot, wearing shorts and T-shirt. He looked so much like Fred Astaire that I expected him to leap into the air, click his heels together, and dance across the beach. He strode toward me, already talking before I could make out his words, gripped my hand and pumped it, still talking, and pulled me up to the cottage to meet his wife, Jean. We had a busy day ahead of us, he said. Lots to do and lots to see and by God we better get our butts in gear. Was I hungry? Did I bring my bathing suit in case we wanted to take a dip? Was I a diver? Did I have a camera or should he get one of his? Let's go, let's go, time's awasting.

First we walked a mile of beach up the Point and Dave talked about prospecting for artifacts—how he goes out in the morning after every storm to see what the waves washed up. This has been his habit for half a century, and he gives no indication that he'll ever quit. Soon after the sun rises he'll fit his 9.5-horsepower Evinrude outboard to the transom of his rowboat, push the boat ahead of him into the waves, and jump aboard as nimbly as a man half his age. He starts the motor with a pull and steers toward the end of the point, following the shore just outside the second sandbar.

Here on the windward shore the water is clear, the beach broad and barren with sand. It could not be more unlike the lush, rich, prolific marshes that crowd the protected lee side, where vegetation grows on land and water both. On the western shore, exposed to the open lake, the waves roll in and shake the sand clean and wash up debris that's been buried for a century or more.

As he motors along, the boat rising and falling on the waves, Dave scans the shore. Cottages soon give way to dunes and woods that have changed little since the glaciers left. If Dave sees something interesting—debris of any kind, any anomalous geometry—he throttles back and studies it with binoculars. If it is a log or stump or some other object of clearly natural origin, he motors on. If it appears to be timber, has an intriguing riblike bend, is black with age, is studded with iron spikes, or is simply unusual, he steers to shore, beaches his boat, and jumps out to investigate.

Often he finds parts of old wrecks. The waters around Long Point contain so many sunken ships that it is a rare storm that does not wash up a section of hull or deck, a piece of transom, a keel, some ribs. In the nineteenth century the remote south shore of the Point was cluttered with such a profusion of wrecks that it became legendary among sailors as a graveyard for ships.

Dave has collected enough artifacts to single-handedly outfit a maritime museum called "Davey Stone's Locker" at the Backus Heritage Conservation Area in Port Rowan. Others clutter the corners and adorn the walls and shelves of his beach house, which has been weathered so long by wind and sand that it looks like an artifact itself. In his backyard is a tepee of beach treasures leaning around a big, silver-trunked aspen. Anchor chains wrap the trunk, the bark starting to grow over them. Sections of masts and old flagstaffs lean there, draped with fisherman's buoys and strings of net floats.

Also in the backyard is a steel canister painted to look like a torpedo and hand-lettered with the words "German U-Boat torpedo captured April 14, 1944." During World War II, Dave was a Royal Canadian Navy sailor on the corvette *Chilliwack*, a small and fast warship that escorted convoys of transport ships across the North Atlantic. Once his ship spent forty-eight hours hunting a U-boat, tracking it until it was forced to surface and was captured. It was the longest submarine hunt in the history of the Canadian Navy.

We drove to a marina where Dave keeps his 22-foot runabout, *Beachcomber*. We got in the boat, backed out of his boathouse, and headed down the canal to Long Point Bay. He had already told me about the Long Point Company, a private duck-hunting club with modest lodge and cabins built on posts in the marshes of Long Point. Now he took me there to meet the caretaker, who walked down to the dock when we pulled up and sat on his heels to chat with us. He and Dave gave a tag-team history of the club. It was founded in 1866 and has counted among its members and guests various Cabots, Lodges, and Morgans, the British royal family, and Theodore Roosevelt. The locals call it "the Millionaires' Club." A couple decades ago a storm from the northeast flooded the marshes and tipped one of the millionaire's cabins into the water, releasing rafts of antique wooden duck decoys worth hundreds of dollars each. To this day you can find them displayed on mantels and windowsills throughout the region.

We cast off and motored to open water. After a few minutes, Dave throttled back and slowed the boat to a troll. We looked over the side at the weed-filled bottom six or eight feet below, hoping to spot the hull of a small sailing vessel Dave had recently discovered there. We couldn't find it, and he wondered if we were quite far enough north.

The bay contains a few wrecks, but nothing like the numbers found on the open, western shore of the Point. They're stacked on top of each other out there, some of them old victims of the "blackbirders" who prowled Long Point in the nineteenth century. In those days of few lighthouses it was customary on stormy nights for people to build bonfires to guide sailors around dangerous headlands or into harbors. Blackbirders built bonfires too, but they built them far short of safe passage. Sailors would see the lights and breathe a little easier and steer east of them, figuring they were rounding the Point. Instead, they ran aground on the shoals. The crew would swim ashore, or sometimes not, in either case receiving no assistance from those waiting on the beach. The vessel would be pounded to pieces in the surf and the blackbirders would go to work salvaging the cargo and transporting it down the beach by wagon, selling it to merchants in towns along the coast.

Even the honest people of Long Point earned extra money salvaging wrecks. Often it was lumber or coal, but at least once the cargo was whiskey. During Prohibition, from 1920 to 1933, enterprising smugglers crossed constantly between Canada and the United States. Long Point, the narrowest spot on Lake Erie, became a favorite crossing place. Canadian fishermen and pleasure boaters would transport loads of liquor to remote coves along the Point, where they would rendezvous with boats from the United States. When the Coast Guard caught on and started making arrests, the smugglers purchased thirty- and forty-foot Jardines and Bancrofts, armored them with bulletproof plating, and modified them with 500-horsepower Liberty Aircraft engines. These sleek "no names" and "grey ghosts" could outrun any Coast Guard vessel and were impervious to machine-gun bullets. The Coast Guard countered with a 42-foot Jardine of their own. Equipped with three 500-horsepower engines, it was capable of speeds over sixty miles per hour.

In a November storm in 1922, the steamer *City of Dresden* ran aground off Long Point while carrying $65,000 worth of illegal whiskey. Some of the cases washed ashore and residents hauled them away as fast as they could gather them. Knowing that revenue agents would soon arrive, they buried the booze in their gardens and honey holes, hid it in their haylofts, cellars, and attics. Dave remembers not many years ago when someone demolishing an old house in Port Rowan found a dusty bottle of whiskey secreted in the wall. It was said to have aged deliciously.

Dave was an early enthusiast of the scuba gear pioneered in the 1940s by Jacques Cousteau and others. While still a young man, he began diving on the shoals around Long Point. He found wrecks beyond belief, including more than thirty that had not previously been documented, and eventually mapped the resting places of nearly two hundred. He and his friend and co-author, Dr. David Frew of Gannon University, counted 429 wrecks in a broader area they call the "Lake Erie Quadrangle"—2,500 square miles centered on Long Point on the north shore and Erie, Pennsylvania, on the south. In their book, *Waters of Repose: The Lake Erie Quadrangle,* Stone and Frew compare this concentration of wrecks to the South Atlantic's Bermuda Triangle, which is five times larger than the Quadrangle but has been the site of only 112 wrecks. According to their calculations, the Lake Erie Quadrangle is not only the "graveyard of the Great Lakes," but the most dangerous water in the world.

Every kind of vessel has been lost there, from the sloop *Detroit,* which foundered near the New York shore in 1797, to the freighter *James Reed,* which sank off Long Point in 1944, to the tug *Captain K,* which went down in Long Point Bay in 1991. Not all of them can be identified. Dave and a couple friends discovered a "mystery schooner" in deep water south of Long Point (near the deepest spot in Lake Erie, at 210 feet) that he claims is one of only five sunken schooners ever found with all its masts intact. Diving on it, he said, was like diving on a Hollywood set. It was a ghost ship resting upright on the bottom, everything in place, lacking only skeletons with sabers in their hands.

I asked Dave what he sees when he explores a wreck. It was a loaded question. I'd always imagined diving on a wreck must be like entering a tomb. The stillness and the silence would be in counterpoint to the tumult that must have filled the vessel's last moments before sinking—the wind roaring, the waves exploding over the deck, the passengers and crew screaming. I imagined that seeing a wreck up close underwater would arouse complex emotions, part horror and part perverse fascination, like witnessing a car wreck along a highway.

Dave's response was more optimistic than I expected. "No matter how old the wreck is," he said, "I see a boat that is new. I see it being launched, surrounded by a crowd, maybe a band playing, the owner standing beside it thinking he's going to earn a lot of money from her. A captain, maybe taking charge of his first boat. A young crewmember, maybe about to go to sea for the first time. I don't see just a few old ribs sticking out of the sand, I see the hopes and dreams of a lot of people. And I see a story that will probably never be told."

We passed Long Point in the night. Beyond the blinking lighthouse at its tip lay the eastern basin of Erie. To the south, past the horizon, were a few miles of Pennsylvania shore and the city of Erie with one of the few natural harbors on the lake. A little east of it was New York state, which bordered the rest of Lake Erie and all the length of Lake Ontario.

By now, the crew of the *Malabar* had coalesced. Everyone pitched in without complaint, was flexible about watches, was generally cheerful. I wondered if a trip like this brought out the best in people, or if these were just good people.

Good people, I decided. But not saints. I sat alone in the main cabin with Matt, who was examining charts at one table while I ate a sandwich and read a book at another. "That Tim," Matt said, by way of prelude, and I knew what was coming: ". . . I just don't understand . . . Won't do anything I ask and says I'm on his back all the time, which I'm not, though I should be . . . I can't believe he just walked up and took that line out of Lisa's hand and belayed it himself, as if he knew more than she did . . ."

Later I went below to the galley for a cup of the coffee that always simmered on the back of the stove—and there was Tim in his apron, a smudge of flour on his chin, leaning around me to make sure nobody was in the cabin to overhear, saying, ". . . and if that son of a bitch says anything to me again I'm gonna . . . Can't believe he walked up and grabbed the rope out of my hand like I didn't know how to tie a knot . . . Just keep him away from me . . ."

Monolithic storage towers mark the entrance of the Welland Canal at Port Colborn, Ontario. From far out on the lake we watched the towers grow, then begin miraging until they disconnected from the earth and floated above it.

A few miles beyond Port Colborn is the end of the lake, where all of Erie is pulled into the mouth of the Niagara River at Buffalo. I stopped there during one of my driving tours around the lakes and saw the approximate view La Salle saw as he was launching the *Griffin*. It was impressive. Look westward and the entire body of the lake sweeps toward you, funnels into a bay a mile wide, then narrows in a sudden massive slide into turbulent river. It's like watching the Great Plains get sucked down a drain.

Twenty-two miles down the river is Niagara Falls. According to Hennepin, we should have heard their roar from Lake Erie, but of course Hennepin exaggerated. When he stood atop the falls in 1678, he was perhaps the first white man to see them. Later, when he became the first to describe them in print, he set a tone echoed by many early travelers: ". . . when one looks down into this most dreadful Gulph, one is seized with Horror." How deep the Gulph was a question of some uncertainty to the friar. In the first edition of his memoirs he claimed the falls were five hundred feet high, triple their actual height; for the next edition, he elevated them to six hundred feet. As they would for many of the observers who followed him, the falls filled Hennepin with awe and terror: ". . . a vast and prodigious Cadence of Water which falls down after a surprising and Astonishing manner, insomuch that the Universe does not afford its Parallel. . . . The Waters which fall from this horrible Precipice, do foam and Boil after the most hideous manner imaginable, making an outrageous Noise, more terrible than that of Thunder; for when the wind blows from the South, their dismal roaring may be heard above fifteen leagues off. . . ."

Niagara makes a formidable barrier. For ten thousand years the falls kept the upper lakes biologically isolated, keeping Atlantic salmon out, for example, though they thrived in Lake Ontario until the late nineteenth century. For the original human inhabitants and several generations of Europeans, Niagara forced a long and difficult portage from Lake Ontario to Lake Erie.

Port Colburn has a clean, modern, welcoming marina with showers, a Laundromat, and a tavern on the premises. We moored at the public dock, throwing our lines to people who had gathered there as we approached. Two of them, a married couple, came up shyly after the boat was secured and said they had taken a day cruise on the *Malabar* during their honeymoon in the Virgin Islands in 1981, when the boat was still called *Rachel Ebenezer*. All of us on board wished she had kept that lovely name. *Malabar* had long been used for a series of boats owned by famed shipbuilder John Alden and it gave off an aura of prior claim. Still, the married couple confirmed our growing regard for the boat. She had the wrong name but was memorable for the right reasons, and, as a bonus, had always made money for her owners—a rare quality among tall ships. Hajo wanted to buy her.

After showers, laundry, and dinner, we sat on deck enjoying the evening air and the day-off feeling that comes with not having to do bilge checks and night watch. It was good to be clean, warm, dry, and dressed in fresh clothes. We took stock of the journey so far.

"When we started," Hajo said, "this was just a delivery, just a job. But now it's really cool. Every step is a new challenge. You get one challenge out of the way, you face another, then another."

I asked him what he thought of the Great Lakes now. "I'm surprised at how much seamanship is required to sail them. I always thought they were for wussies, that only the oceans were worthy of tough guys like me. But in the ocean there's not much to hit, it mostly requires endurance. These lakes can kick your ass."

. . .

We entered the Welland Canal at ten-thirty Friday morning. First order of business was to pass customs, which on this sliver of Canada meant chatting with a customs officer as he walked along the bank beside us. "Any alcohol on board? Any firearms?"

No and no, said Hajo. But he forgot to mention that we had hidden in devious cubbies below deck the dozen cases of homemade beer and some bottles of rum and single-malt Scotch and two or three jugs of wine. Also a cannon. It was Matt's delight—a lovely brass swivel gun that mounted on the gunwale and was bored to shoot a ball an inch in diameter. With it were a couple pounds of black powder, some wadding, an assortment of arcane tools for loading and cleaning. No cannonballs, though. We would fire blanks.

The customs agent studied Hajo's face for traces of perfidy. Hajo met his eyes and grinned. Who could resist that smile? Those clear blue eyes? Those sea-captain muttonchops? The agent nodded and walked off to interrogate the next group of boaters. Hajo winked at me. We felt like pirates.

The locks of the Welland Canal bypass the 326-foot descent of the Niagara River by channeling it into eight navigable steps. First comes a "regulating lock" only a couple feet high, to adjust for fluctuations in Lake Erie, followed by several miles of canal passing through pleasant country, wooded and rolling. Then comes the first lift lock, which we entered through an open gate to a rectangular chamber large enough to accommodate an oceangoing ship. Two small sailboats entered after us and sulked in opposite corners.

Friendly attendants strolled to the edge of the lock to chat. Where were we from? Where were we headed? These attendants possess what must be among the most coveted government jobs in Canada. Their primary duty is to be ready on the top of the locks so they can toss lines to boaters as they enter, one line fore and one line aft. Between boats they sit in lawn chairs beneath awnings in front of the cottagelike lockmasters' offices, where they keep charcoal grills burning and coolers of soft drinks cooling.

They threw us our lines. "Hold 'em, don't tie 'em," they said. Good advice. The idea is to let the lines play out as the water in the chamber drains and your boat descends. There would be some turbulence, so the lines served to keep us from swinging away from the walls and possibly crashing into other boats in the lock.

Because we didn't want to be mashed against the walls, we had followed the official recommendation and built a pair of fender boards to hang against the hull. Each was constructed of an eight-foot length of two-by-eight lumber with a pair of old tires lashed to it. We hefted them over the side, one near the bow and one near the stern, and belayed them in place. For extra cushioning we had purchased a half-dozen plastic feedbags stuffed with straw and tied to a short length of rope. At five dollars each, they were a bargain. They and our "hippity hop"—a large orange rubber ball on a rope—would prove indispensable.

The massive gates swung closed behind us. Slowly, silently, the level of the water began to fall and we descended. About twenty-one million gallons were draining through sluices in the bottom of the lock, carrying us down to the next level.

A lock works on a basic principle: water craves equilibrium. The simplest lock is composed of two chambers arranged end to end, each sealed with watertight doors. If a boat is headed downstream, as we were in the *Malabar*, it enters the open gate into the first chamber, the gate closes behind, a valve is opened, and water drains into the chamber below. As the higher chamber empties, the lower one fills, until they reach the same level. The door between them is then opened, and the vessel proceeds on its way. For a boat headed upstream, the process is reversed: the vessel enters the lower chamber, the gate behind it closes, water from above pours in, and the water in the chamber rises to meet the level of the chamber above.

Although some locks are closed at their ends with vertical gates (we would pass through one on the Erie Canal), their overhead gantries and counterweights restrict the passage of vessels with masts or superstructures. Most locks, those in the Welland included, use swinging, double-leaf doors called mitre gates. Mitre gates cannot be opened or closed until the water pressure is equalized on both sides, thus forming a strong and waterproof seal. Some historians are convinced that mitre gates like those used today were invented by Leonardo da Vinci for the San Marco Lock in Milan.

We descended forty-five feet in about ten minutes, until we were deep inside the chamber, looking up at towering cement walls dripping with water. It was cool and damp and shadowy down there, like being in a river canyon. Matt and I hooted to hear our echoes.

The water level inside our chamber reached equilibrium with the canal below, and the lower gate opened. We tossed the mooring lines away, and they were instantly hauled up by the attendants, who leaned out high above us and waved. The little sailboats shot ahead. A couple others were waiting to enter after we left; they would ride the water up as the lock chamber refilled. We idled into the canal and went on to the next lock and repeated the process.

In the long history of the Welland Canal, hundreds and probably thousands of schooners not much different from the *Malabar* made this trip. A typical one was the *William Sanderson*, a schooner built in 1853 in Oswego, New York, on the shore of Lake Ontario. She was designed to be a "forwarder," a vessel that could carry wheat and other grains east from Chicago to Oswego, where the cargo was offloaded to barges and shipped through the Erie Canal to markets in New York. She would then be loaded with lumber and household goods and would return west to Chicago or other lake ports. The *Sanderson* had a capacity of 20,000 bushels and was built to the dimensions of the second Welland Canal, opened in 1848—136 feet long and with a beam of slightly over 25 feet.

Wooden sailing vessels of that era usually lasted only about fifteen years. Many sank or were wrecked, but most were used until their hulls deteriorated and became so unsafe that crews refused to sail on them. They were often dismantled for scrap or stripped down and converted into barges.

The *Sanderson* suffered her share of mishaps. In November 1857, she ran aground in Lake Erie near Buffalo and had to jettison a portion of her cargo of wheat before she could be floated again. A more serious accident occurred in the Straits of Mackinac in November 1874, when her captain was knocked overboard by the main boom and drowned. The crew continued on to Chicago, where another captain was hired (his previous vessel had just sunk near Sheboygan, Wisconsin). The new captain guided the *Sanderson* out of Chicago on November 20, carrying a cargo of 19,500 bushels of wheat, bound for Oswego. Six days later her demolished hull washed up on the Michigan shore near Empire, just south of the Sleeping Bear Dunes. None of her crew was ever found.

Canals and locks are integral to the St. Lawrence Seaway, the mammoth shipping project that opened the Great Lakes to oceangoing traffic in 1959. Without the locks around the rapids of the St. Lawrence River, in the Welland Canal, and at Sault Ste. Marie, there could be no significant shipping in the North American interior.

Construction of the original Welland Canal was begun by Canada in 1825 to compete with the Erie Canal, which the government worried would lure lake commerce southward. It was completed in 1829 and consisted of forty locks big enough for schooners up to 110 feet long. The *Malabar* would have been about as big as any boat that used it. In the years ahead the canal and its locks would be enlarged many times, until by 1932 the locks reached their current size. Each chamber is eight hundred feet long and eighty feet wide and deep enough for vessels drafting twenty-seven feet.

The opening of the Welland Canal, and especially its subsequent enlargements, allowed ships from all over the world to come to the upper lakes, fueling the growth of Cleveland, Detroit, Chicago, and most of the other port cities. But nobody could have foreseen that the canal would also allow entry to a most unwelcome visitor, the sea lamprey.

It's difficult to exaggerate the impact of this invader. By the beginning of the twentieth century, the Great Lakes were the richest freshwater fishery in the world. In good years commercial nets hauled out millions of pounds of lake trout, whitefish, chub, perch, and walleye. But those good years were soon gone.

The sea lamprey is a primitive, snakelike, jawless ocean fish often mistaken for an eel. It feeds by attaching to other fish with a sucking disk around its mouth and uses its sharp teeth and tongue to rasp through scales and skin until it strikes body fluid. Though only about a foot in length, each sea lamprey can kill forty or more pounds of fish in its lifetime. Only one of seven fish it attacks is likely to survive.

Sea lamprey were seen in Lake Ontario as early as the 1830s and were probably always present, their population held in check by natural factors, including too-cold water in most of the tributaries where they tried to spawn. But after the logging of the region in the late nineteenth century, the rivers were no longer shaded and became much warmer. This made them unsuitable for Atlantic salmon, which disappeared from the lake, and very suitable for lamprey, which flourished in it.

Lamprey might previously have made their way into Lake Erie in small numbers, but the invasion did not begin in earnest until the Welland Canal was widened and deepened in 1919. From then on, lamprey marched steadily through the lakes. They were observed in Lake Erie in 1921, in Lake St. Clair in 1934, in Lake Michigan in 1936, and in Lakes Huron and Superior in 1937. As they colonized the lakes, they annihilated lake trout, whitefish, chubs, burbot, walleye, and catfish, as well as rainbow trout and other introduced species. In Lake Huron, the annual harvest of lake trout fell from 3.4 million pounds in 1937 to virtually nothing in 1947. Lake Michigan's catch plummeted from 5.5 million pounds in 1946 to 402 pounds in 1947. As many as eighty-five percent of the fish caught and netted in those years were scarred with lamprey wounds. Then the fish were gone. Hundreds of commercial fishermen, many of them from families that had been working on the lakes for generations, were ruined.

In the 1950s, scientists under the direction of the U.S. Fish and Wildlife Service declared war. First they had to learn everything they could about the lamprey. They discovered that in spring and fall the adults swam up tributary rivers to spawn in gravel riffles. About ten percent of the 5,747 streams that flow into the Great Lakes were found to be used by lamprey for spawning. The adults built nests in the gravel, laid eggs, and died. The eggs hatched into wormlike larvae that drifted downstream and burrowed into sand and silt, where they spent the next three to seventeen years feeding on algae and bottom debris. When they grew to about six inches long, they transformed from the larval stage to the parasitic stage and migrated to the open lakes. There they spent the next twelve to twenty months feeding on fish before returning to the streams to spawn. The total life cycle averaged six years, but individuals might live as long as twenty years.

Young lamprey still in their natal streams were more vulnerable than adults in the lakes, so biologists focused their attention on ways to poison them. After testing nearly six thousand chemical compounds, in 1958 they isolated one that killed sea lamprey larvae without harming most other organisms. They began dispensing the compound, TFM (3-trifluoromethyl-4-nitrophenol), into tributary streams, with immediate results. Within a few years the sea lamprey population in most parts of the Great Lakes was reduced ninety percent from the levels of the 1940s and 1950s, and the way was cleared to rebuild the fisheries.

Rebuilding in Lake Erie meant first cleaning the water; that was challenge enough. The other lakes, however, faced different challenges. Though their major bays were seriously polluted, the bodies of the lakes remained in good shape and should have supported healthy fisheries. But dominos were falling. The sea lamprey had wiped out most of the trout, whitefish, and other predator fish, which allowed the population of alewives to explode.

Alewives are ocean fish that invaded the lakes about the same time as the lamprey, but at four to six inches in length they were too small to be prey. Once in the lakes they reproduced prodigiously, outcompeting native species for food and filling their niches in the ecosystem. By the time the sea lamprey was brought under control in the early 1960s, eighty-five to ninety percent of the biomass of the lakes was made up of alewives. They virtually filled the lakes. Periodically they died off, millions of them at a time washing to shore and fouling beaches.

When I was a kid, we could smell the alewives a mile away from Lake Michigan. Waves had shoveled them into waist-high piles along the shores, where they rotted. They were a disaster for tourist industries already hurt by pollution and the collapse of the sport fishery. In desperation, people sought ways to get rid of them. Commercial fishermen studied the feasibility of harvesting alewives for cat food and even invited fishermen from the East Coast to bring their trawlers to the lakes. But biologists had another idea. Instead of a nuisance, they saw alewives as food for other fish. All they needed was a predator.

Since early in the twentieth century, fisheries managers had tried to introduce Pacific and Atlantic salmon into the Great Lakes. Those efforts had always failed, probably because immature salmon fry were planted instead of larger and hardier smolts. This time, Michigan Department of Natural Resources biologists, led by Dr. Howard Tanner, head of the Fish Division, and his assistant, Dr. Wayne Tody, took a bolder approach. Although aware of the gamble—nobody could be certain that Pacific salmon would adapt to a freshwater environment—they obtained a million coho salmon eggs from Oregon and placed them in state trout hatcheries. After a year and a half of care, 850,000 healthy smolts were produced. In the spring of 1966, 650,000 of them were released into Michigan's Platte and Manistee rivers, both of which flow into northeast Lake Michigan.

Not even the most optimistic biologists were prepared for what happened next. By the end of the summer of 1967, smolts that had weighed about an ounce when they were released sixteen months earlier now weighed ten to twenty pounds. Immense schools of them gathered off the mouths of the Platte and Manistee rivers, and anglers from all over the United States converged. The gamble had paid off. Other states quickly initiated salmon programs of their own and began releasing coho and the larger Chinook (or king) salmon into all five lakes. Within a few years the recreational fishery for Great Lakes salmon was one of the most vital in North America, and the alewife problem was under control.

Today, several hundred streams are treated with lampricide at intervals of three to ten years each, and others are blocked with weirs, electric barriers, and other control measures, but the balance is tenuous and lamprey remain a constant threat. Their control is so expensive that even a small cut in the budget could be disastrous. Currently, the most serious problem spot in the Great Lakes is the St. Mary's River between Lakes Superior and Huron. Millions of lamprey spawn there, but the river is so big that treating it with lampricide has never been practical. So, while lamprey numbers have declined in most parts of the Great Lakes, they've been increasing again in northern Lakes Huron and Michigan. But for now the fishing remains very good.

Chapter 12

LAKE MICHIGAN

Lake Squall, 1967

I was thirteen the year the Great Lakes came alive again. My family and I lived a few miles from Lake Michigan, on the shore of an inland lake where I should have found plenty of opportunities for adventure. But I was impatient. I longed for uncommon experience. When storms chased vacationers inside their cabins, I took the rowboat out on Long Lake or pestered my parents until they drove me to Lake Michigan so I could stand in the wind and watch waves. I scanned the sky for thunderheads and twisters. I wanted a life filled with drama. And then one day I learned about drama.

The salmon that had been imported as eggs from Oregon thrived in Lake Michigan. They found outrageously abundant forage there—alewives, tons of them—plenty of space, and no competition. They grew at an unprecedented rate. By August 1967, people in aircraft saw schools of salmon three or four miles long swimming toward Platte Bay. The word spread.

A kind of gold rush mentality prevailed. Anglers drove straight through from Pennsylvania and Tennessee and North Dakota. Most had no idea how to fish for salmon. Many had never before visited the Great Lakes and were surprised they couldn't see from one shore to the other. Towing the same small boats they used back home for bass and panfish, they got in line at the access ramps at the mouth of the Platte—the queue of vehicles stretching three miles up the road. When their turns came, they backed rapidly down the ramp, hit the brakes, and catapulted their boats into the river. Often in their frenzy to fish they parked their cars and trailers in loose sand beside the road, burying them to the axles and abandoning them. At midnight they would still be there with flashlights, digging with their hands and jamming scraps of plywood and carpet under the tires. But that was later. Now they ran to their boats, yanked the starter cords, and were planing at top speed when they reached the gravel shoal where the river meets the open lake.

Out in the bay, boats trolled back and forth in confused clots. As many as fifteen hundred vessels at once converged there. Every type of boat could be seen, from fourteen-foot runabouts to forty-foot yachts, from twelve-foot prams to thirty-foot sailboats. My father and I watched in amazement as a fisherman, five miles from shore, paddled past in an aluminum canoe. Once we saw two men crammed inside an eight-foot dinghy that looked more like a bathtub than a boat. It was common to see six adults crowded in a tiny boat with four inches of freeboard.

When the water was calm, any boat could handle the lake and any angler who figured out a way to drag a bright lure through the water could catch salmon. The fish averaged ten to fifteen pounds each and were as bright as silver ingots. And like precious metals they inspired a lust for acquisition. People became greedy. They went out in the morning, caught a limit of two salmon per angler, motored to shore to put the fish on ice, then returned to the lake and caught another limit. It was terrific fun. A carnival atmosphere. Now and then tempers flared when someone tried to go out of turn at the access site or if fishing lines from two or three boats got tangled, but spirits ran high and people laughed and shook their heads in wonder at the bizarre things that happened. Salmon "porpoised" from the water, leaping so high they sometimes landed in the laps of surprised anglers. A dozen or twenty boats would troll through a school and everyone with a line in the water would hook a fish at the same time. Sometimes, while you brought one salmon to the net, two or three others would follow and slash at the lure in its mouth.

No one had seen anything like it. This lovely vast lake, blue and limpid and bordered by rolling dunes, had been depleted of gamefish for so long that few anglers even considered fishing there. It was a stinking mess in the summer when the alewives died and a scenic desert the rest of the year.

But now the lake produced unbelievable bounty. The local economy boomed overnight. You could search three counties and not find a vacant cabin or motel room. Every store was cleaned out of its entire stock of fishing lures. Line, sinkers, rods and reels, landing nets, coolers—all were sold out. It was a happy, manic, innocent time.

We had coho fever like everyone else. Saturday and Sunday mornings my father, mother, brother, and I trailered our boat to the mouth of the Platte and got in line at the access ramps. Our fiberglass runabout was only a fourteen-footer but it was deep-hulled and more seaworthy than most, with a canvas top, and an outboard motor powerful enough to pull a skier and a small trolling motor my father rigged to the transom for backup. Usually we were on the lake before dawn, our rods set and lures in the water when the sky grew bright. We fished with the same spinning rods we'd always used, with reels that held a hundred yards of eight-pound monofilament line, and for lures used Flatfish, Daredevls, and Rapalas—the same lures we used for pike and bass.

I remember a day that might have been our first attempt to catch salmon. My father and mother, my brother Rick and I were together in the boat, all of us excited and alert. I remember the lake was calm, but big gentle swells passed beneath us as we motored out of the river and away from shore. I remember the smell of engine exhaust mingled with fish scent, seeing dozens of boats in clusters not far offshore from the river. The lake seemed larger than I'd expected and our boat smaller. We were equipped with our rods and tackle boxes and a cooler full of sandwiches, and were determined to catch the exotic new fish we'd be hearing so much about.

Later we would learn that the best way to find salmon was to follow crowds of boats. We would scan the water with binoculars, and when we saw congested boats and bent rods, would speed close and cast our lines out and begin trolling. But my father preferred to fish in solitude, and those first few times we went our own way.

We motored two miles, three miles, four miles offshore, until we were farther out than anyone else. From that distance the shore was a low border of green against the lake and the water beneath us was deep blue and so clear you could look down and down and imagine you could see forever.

My brother and I rode in the open bow of the boat, scanning the water ahead. I saw a swirl on the surface. Then another. And another. A large fish as bright as stainless steel vaulted into the air and landed in a blossom of spray.

Salmon!

Dad cut the motor, and the boat heaved to a stop. We were surrounded by silence. I looked down into the water and saw blue streaks firing past. There were many of them and they were very fast. They looked like bolts of ice-colored electricity shooting by, a river of them flowing rapidly around us, as if we were a rock in midstream. We were in the midst of a gigantic school of salmon. For a hundred yards in every direction they swirled and porpoised and leaped free of the water. They were feeding on alewives, which tried to escape by skittering across the surface. It was a massacre.

"Cast! Cast!"

We grabbed our spinning rods and cast wildly. Fish slammed the lures, first one then another. Mine leaped, and ran beneath the boat and leaped again—so fast that I thought it had to be a different fish. How could a single fish get from there to there in half a second? Then it tore line off my reel in a lunatic run. It went a hundred feet, two hundred feet in the time it took me to shout for help. It was unbelievably strong. I didn't know any fish could be that strong or could swim that fast. I was never in control of it. But somehow I got lucky and the fish stayed on my line. Dad netted his, a brilliant ten-pound coho, then mine, almost identical, and we cast again and Rick caught one and my mother caught one and my father another. Then the school disappeared and we went hunting for others. The experience had been amazing and magical and we were convinced that anything could happen on the suddenly revitalized lake. We were hooked more deeply than the fish we caught.

During the next two weeks we caught dozens. Everyone did. Every time we went out on the lake, we came back with our cooler filled. We took the salmon home and filleted them or cut them into steaks and ate them baked, grilled, broiled, and fried. What we couldn't eat we froze or canned or gave away to friends.

For those weeks the wind stayed light, the sky was clear, the days were warm and pleasant. It was easy to believe that it would stay like that forever. But my parents were concerned. They knew the lake would change.

Mark Dilts, a columnist for the *Traverse City Record-Eagle*, shared their concern. He wrote on September 2: "When the big lake is calm you can paddle around in it in a washtub without any difficulty. But when she begins to kick up her heels she can be a holy terror, making brave men pray for their deliverance from her clutches. . . . Lake Michigan, like a beautiful lady, commands respect from all who know her. For those who have yet to learn this, the lesson can be a costly one. It can even be fatal."

Meteorologists sometimes describe the Great Lakes as weather factories. The five lakes are so large and contain such a tremendous volume of water that they modify the climate of much of the center of North America. And because they straddle the 45th parallel, halfway between the North Pole and the equator, the sun's angle of inclination and the length of the days varies greatly as the seasons progress, making winters cold and summers relatively warm. Furthermore, the lakes are located near the center of a vast, wedge-shaped trough cutting north and south across the middle of the continent. This lowland trough allows cold, dry air from the Canadian north to meet warm, moisture-laden air from the Gulf of Mexico. With no mountain ranges to block their passage, the air masses come and go, driven by the wavelike undulations of the jet stream, producing hot weather when the jet stream loops north and cold weather when it loops south.

All these factors contribute to make the Great Lakes an extraordinarily active weather zone. Locals have more justification here than in most places to say that if you don't like the weather, wait five minutes. Temperatures can swing forty degrees in a few hours. Cold fronts sweep down from the north with almost no warning, driving warm, humid air upward, forming clouds and precipitation. Violent weather often results.

Autumn is an especially unstable season around the Great Lakes. The surface water is often warmer than the air, so as cool wind from the land passes over the lakes, it picks up heat and moisture. The warming air be-

comes more buoyant and rises, carrying moisture with it and creating turbulent winds and isolated storms. Larger systems of storms surge east across
the Great Plains. When they encounter the lakes, they intensify. By the
time they cross to the opposite shores, they can be monsters.

That Saturday of the autumn equinox, September 23, 1967, promised to
be another fine day. Friday's forecast gave no indication of trouble: "Saturday partly cloudy and a little warmer, with a chance of showers near
evening. Northerly winds . . . light and variable." For weeks, newspapers
had published newswire reports raving about the spectacular fishing in
Platte Bay. It was on the minds of anglers all over the country. Thousands
of them finished work Friday, loaded their boats, and drove north.

But early Saturday, the weather took an unexpected turn. A shift in the
jet stream had brought a sudden cold front down from Canada. At four-
thirty that morning, when the alarm went off, I could hear the wind in the
trees. My mother argued for staying home. Dad and I talked her into giving
it a try.

At dawn we stood beside the tiny weather shack near the mouth of the
river. A single red pennant snapped in the wind above us. Small-craft warnings. The phrase had potency to anyone who knew the lakes. My father
had spent enough time on them to be wary. My mother was even more
cautious. She had grown up on the shore a few miles north of Platte Bay,
and had spent her childhood walking the beaches and listening to stories
of storms and shipwrecks told by her father, who had been a crewman at
the Coast Guard Station on South Manitou Island. We had all witnessed
storms, had watched gigantic waves batter breakwalls and lighthouses and
send spray bursting twenty or thirty feet in the air. The shoals around the
Manitou and Fox islands had wrecked dozens of ships. Sometimes after
storms we found artifacts on the beach—the waterworn ribs and keels of
old sailing ships, each a reminder of the lake's power and our frailty. I had
been out with my father in three-foot waves so choppy they seemed about
to break our boat to pieces. Even much larger craft were at risk when the
waves exceeded four or five feet. They were not regular swells. They were
tight and steep and unpredictable. They could break a boat in half.

To the west, the sky was dark with squall lines. Already the lake was the color of steel and booming with whitecaps. During the drive to the Platte, before daylight, we had listened to weather reports announcing waves two to three feet high and winds of twenty-five knots and increasing. It was hard to ignore the warnings.

Many anglers ignored them anyway. Those who had traveled long distances were especially reluctant to stay on shore. After driving all night to reach the lake, they refused to be stopped by a few waves. Besides, others were going out, hundreds of them, and there was comfort in numbers. The general assumption was that small-craft warnings were a formality, a way the Coast Guard and the National Weather Service could avoid liability should someone run into trouble. Only one thing was certain: You couldn't catch fish on shore.

So they went out. The Coast Guard later estimated that more than a thousand boats motored into the waves beneath that moiling black sky. My father and I watched them. We stood on shore with the wind in our faces, smelling the big-water smell of the lake and the stench of the salmon guts spilling from garbage cans in the parking lot. We watched anglers one after another launch their boats and motor down the estuary at the end of the river until they met the breakers at the mouth. Their bows rose toward the sky and fell, like military tanks busting over berms. The waves struck the boats and sent up spouts of spray that were quickly sheared off and blown toward shore. A few boats turned back at the mouth and retreated upriver, their passengers shaking their heads in defeat. But for every one that returned there were a dozen waiting to challenge the waves.

All morning conditions worsened. By afternoon, the wind had reached forty miles per hour and the waves were six to eight feet high. The air was filled with the roar of wind. Yet anglers were not deterred. Hundreds of them stayed on the lake and fished. Then their boats began to swamp.

At first they tried to reach shore by motoring through the ranks of breakers at the mouth of the river. The only way past the gravel bar was through a single narrow channel where the strongest current flowed. It was tricky, even in calm water. In the high waves, boats came in from all directions at once and wedged in the channel. They rammed one another, turned sideways in the current, swamped when waves broke over them. Soon dozens of boats were engulfed.

More timid boaters stood offshore, hoping the wind would change and the seas diminish. They circled out there, fighting the waves until their fear of the open lake overwhelmed their fear of the breakers and they made a dash toward the beach, adjusting their throttles to the waves in an effort to run aground between them. They came in, their engines laboring one moment and screaming the next as the water fell beneath them. Six or eight of us on shore would run down with the descending wash and grab the boats by their gunwales and docking lines. The men inside jumped out to help and the women and children crouched on the decks with terrible looks on their faces, and we would pull the boats as far as we could up the streaming slope of the beach before they were slammed by the next wave. We were successful only with small boats. Larger ones were too heavy to pull. We would hold them as best as we could while they wallowed in the surge until a wave washed over their sterns, dumping a roiling froth of water and sand inside. A few waves later the boats would be awash in the surf or anchored to the bottom with sand.

Rumors ran up and down the beach. Hundreds were missing, presumed dead. Bodies washed up at Frankfort. Dead children at Empire. Boats lost far out in the bay, beyond help.

Days later, we would learn that most of the missing had been accounted for and that only seven men had died. It was a wonder. One rescuer said that of the fifteen or twenty boats he helped drag onto the beach at Empire, only two contained life preservers.

There were heroics. Coast Guard helicopters lowered baskets to floundering anglers and lifted six of them to safety. Two men clung to the side of their capsized boat for more than two hours until they lost consciousness in the fifty-degree water and were rescued somehow by people on shore who waded through the surf and dragged them to safety.

My father and I helped as much as we could. Dad had been a police officer and was trained to save lives. I knew he could rescue anyone in danger. The knowledge was exhilarating. It made me more competent just to be with him. I felt capable of adult heroics.

Somebody told us people were in trouble at the boat ramp in Empire. We drove there to help and joined a small crowd on the beach. Behind us was the house where my family had lived the summer I was five years old. Before us was the lake, gone insane with whitecaps and breakers. A few abandoned boats had washed up on shore and were full of sand and water. A boat with two men inside circled beyond the breakers. The men seemed unsure of themselves. They had watched others try without success to run the gauntlet of breaking waves and seemed to be looking for a way to save their boat. They circled, rising on each wave, disappearing into each trough, their heads swiveling as their boat turned, always facing shore. You could see them working up their courage.

Finally, they steered toward the beach. Instead of accelerating like most of the others, they came cautiously, their engine at trolling speed. They went up on a wave, down in a trough, up on another wave. They went down in a trough and did not come up. When the wave passed, the two men were in the water.

They were so close to shore we could see the hair plastered to their scalps and could see the expressions on their faces. They looked more surprised than frightened. Their eyes were big and they worked their mouths, as if apologizing. They bobbed low in the water in their orange life preservers. Every time a wave came over them, they disappeared for a few moments in the froth.

Waves broke on shore with so much force the ground shuddered. I stood on the beach thirty or forty feet above the wash and felt the booming *thump* of every wave under my feet. My shoes were soaked with water and full of sand and my socks had fallen down around my heels. I was amazed that I could feel the impact of the waves through the earth. A simple seismological fact, but it filled me with wonder.

The breakers shoved the men toward shore, then dragged them away again. They never got closer. The current pulled them down the beach away from us. We walked beside them, shielding our eyes from the spray and sand thrown at us by the wind. Every time a wave broke over the men, they tumbled in the foam. Sometimes they turned upside down and kicked their legs in the air as if trying to run. The wave would pass and they would struggle upright. The troughs between the waves were calm, and allowed them to get a few breaths before the next breaker came. They tried to swim but all they could do was splash with their arms. The life jackets seemed to inhibit them more than help them.

People on shore ran to the water's edge carrying coils of rope and tried to throw them. The ropes would shoot out and unfurl and hang for a moment in the wind, then come back. One man knotted a rope around his waist and waded into the waves, but he was knocked down, and others pulled him to shore against his will.

Waves broke over the two men, one after another. With each wave they disappeared, and we saw only glimpses of orange in the froth.

I had sand in my eyes. I turned away and rubbed them and turned back and saw the faces of the men in the water. I made eye contact with one of them. He was heavy and gray, the age of my grandfather. He could have been our insurance man or the guy who delivered our bottled gas. He seemed apologetic. I kept expecting him to smile at me and shrug. A wave would crash over him and after a few moments he would come up coughing and spitting water. Every time it happened, he looked a little more apologetic.

A woman standing near me put her hands to her face and screamed for somebody to do something.

Children were excused from responsibility, but I was no longer a child. I was thirteen, nearly fourteen. I was old enough to help. I pitched on a baseball team and could have thrown a rope better than anyone there. I could have heaved it low and hard beneath the wind and made it straighten like a bullwhip and land within reach of first one man, then the other. I was lean and fast and swam well. I could have tied a line around my waist and dived through each wave the moment before it broke and reached the men in the calm of a trough and spoken reassuring words to them as the people on shore pulled us to safety.

A wave broke over them. Their legs rose in the air but did not kick. Another wave came and I could see two dark, slick objects rolling heavily in the spume, waterlogged. My father gripped me high on my arm and turned me away. I tried to look back but he gripped harder and pulled. An ambulance waited in the parking lot, its lights flashing urgently. People ran past, shouting, their voices torn to fragments by the wind.

The men in the water wore bright orange life preservers with bulky collars designed to support their heads above the water. They should have been safe. Everyone said if you wore a life preserver, you were safe. It was an article of faith. Preserver of life. The Coast Guard guaranteed it. Our parents taught us to believe it.

But the heads of the men did not stay above the water. The preservers hadn't worked. The guarantee was not valid.

I had wanted to be a hero.

I had wanted drama in my life.

My father gripped my arm and pulled me across the parking lot past the ambulance, past people holding their faces in their hands. He put me in the car with my brother and mother. They had been there all along, watching. I will never forget the looks on their faces.

For weeks I lay in bed at night hearing the roar of the storm and feeling the awful draining power of the waves. I wanted to remain a child, but it was too late. Childhood fades with the knowledge of peril, and peril is everywhere. My father could not protect me from it. No life preserver could save me.

They died a hundred feet from shore.

Chapter 13

LAKE ONTARIO

The novelty of the Welland Canal soon wore off and its locks became just a slow way to travel—though not as slow, we reminded each other, as portaging around Niagara Falls. At each lock we had to use care to maintain the *Malabar*'s position. If the boat got too close to the wall, we jammed the hippity-hop and the straw bags against the hull to prevent damage. Our other fear was that the stern would swing out, throwing the bowsprit into the wall; or that the bow would swing, crushing the yawlboat.

Hajo was more worried about driving the *Malabar* into a mitre gate. Vessels this large are slow to stop. Once under way, their momentum keeps them going much farther than you expect. Reversing the engine didn't help much on the *Malabar* because the pitch of the propeller—the "wheel" in maritime lexicon—causes the boat to turn abruptly as it backs, regardless of the angle of the rudder, making it difficult to control. We proceeded slowly forward into each lock—so slowly that at the frequent lift-bridges automobile traffic was forced to wait longer than motorists liked. As we inched toward a raised bridge, a lady leaned out the passenger window of a car and shouted, "Hurry up!" A year later, one of those same lift-bridges (the one at Allanburg) would be lowered too quickly, shearing off the wheelhouse and smokestack of a lake carrier. The ship, loaded with 26,000 tons of grain, caught fire and burned, blocking the canal for several days.

At a spot where the canal divides into twin channels, one for upbound traffic and the other for downbound, we met a Liberty Ship headed for Lake Erie. It was the SS *John Brown*, the first Liberty Ship built to haul troops and cargo during World War II, and one of only two (out of a fleet of 2,710) still surviving. That day she was bound for the Toledo Ship Repair Yard to have work done on her hull. Lining her rails were dozens of men, most of them veterans wearing vintage naval uniforms.

Harold went gung ho. He had served in the navy and was now a chief petty officer in the Naval Reserve. In 1961, he served on a destroyer that was sent with two others to a position a few miles off the Cuban coast, ready to provide cannon support during the Bay of Pigs Invasion. At the last minute the mission was aborted, and the destroyers retreated to international waters.

Now he climbed onto the gunwale of the *Malabar* and hailed the Liberty Ship. "Hey, you big gray thing!" he shouted. "Hello, you big beautiful gray thing! I love you! Way to go, guys!"

The veterans cheered and waved.

Harold's eyes were bright. "I'd love to sail with those guys," he said. "But I'm afraid if I did, my wife would kill me. Maybe next year . . ."

Near the end of the canal we entered Locks Four, Five, and Six—the so-called flight locks. Unlike the others, which are separated by lengths of canal, these three lead directly into one another, a trio of big steps down the 150-foot drop of the Niagara Escarpment.

Eight hours after entering the canal and 27.6 miles from Lake Erie, we entered Lake Ontario. Its surface was calm, painted with evening pastels. The Toronto skyline was clustered on the far shore, the CN Tower thrusting upward like a distant Byzantine steeple.

At 7,340 square miles, Lake Ontario is by far the smallest of the Great Lakes. Though it widens to 53 miles halfway down its 193-mile length, the western third is small enough that you can see both shores at once. The land that surrounds it is so low and distant that the lake appears slightly domed, as if you were sailing atop an immense contact lens.

But though Lake Ontario has the least surface area of any Great Lake, its maximum depth of 802 feet is exceeded only by Superior's 1,333 feet and Michigan's 923 feet. With its surface just 243 feet above the Atlantic, more of Lake Ontario sits below sea level than above it.

While the upper lakes had been plagued for years by low water, local rains kept Lake Ontario at normal levels. Now it was a little higher than normal. The storm that had glanced against us on Erie had blasted the New York shore with floods and wind damage. Rivers still spewed plumes of brown water into the lake, filling it with debris churned to pieces by rapids and falls. We had to keep constant watch. The bowman would point at obstacles ahead and Hajo would steer in zigzags around them. There were logs and stumps to avoid, and entire trees nine-tenths submerged. A few times we plowed through acres of smaller debris.

By dark, we had passed beyond most of the debris and could relax a little. Harold scanned the water ahead with a spotlight, but Hajo and Matt decided it was no longer necessary to be so vigilant. The remainder of the night we motored eastward without incident.

Again I took the graveyard shift. By now I looked forward to the night hours. I loved the way time expanded then, the hours billowing away in every direction. It was the best time to be awake.

That night Hajo and I sat quietly beside the helm, lost in thought, watching the Milky Way above us and the horizon-glow of cities in the distance. After a while we talked. We covered the human condition and touched on various mysteries of the universe. I suspect we were brilliant. Most of the words drifted off and were lost among the stars, but now and then I made notes. Once Hajo said, "A clock ticking in a room is just a reminder of time passing, but you ever notice that a roomful of clocks is music? The same with water. A single drop dripping over and over can drive you crazy, but put a billion drops in a river and you have music."

We talked about how people crave a quieter, slower life but have no idea how to achieve it. For many, life is a competition, and we're not willing to go to the sidelines, even briefly. We're afraid we'll be passed over, forgotten, replaced, made obsolete. We forget Ecclesiastes or deliberately misread it and conclude that the race *is* to the swift. Besides, we like a fast pace. It's intoxicating. It releases endorphins. We achieve a lifestyle equivalent of runner's high. Even as we lie awake in bed buzzing, our veins flowing with caffeine and adrenaline, we congratulate ourselves for keeping the pace. We like the rush. We like talking on a cell phone while signing a contract and reading e-mail and watching CNN on TV, headlines scrolling along the bottom of the screen. We like the satisfaction of making split-second decisions while driving seventy-five miles per hour through heavy traffic. It's problem solving at the highest level. It takes skill. It requires boldness and agility. To roar at maximum speed through the world while the less fit crash behind you is strangely gratifying.

But what if it does damage? What if the pace wears insidiously? Maybe it beats us down. Maybe it dulls our sensibilities and makes our souls withdraw. Maybe we don't notice the damage until we step away from our normal life and experience a slower, quieter, less demanding one. That's probably why we seem to spend the first three days of every vacation decompressing. Not until the fourth day do we remember how to relax.

Living on the boat made all that seem clear. Hajo said he'd noticed a change in me. After only a week on board, modern life was sloughing away from me, like old skin, and I was rediscovering patience. During the Chicago-to-Mackinac Race I had been too busy to think about it, but on the *Malabar* I had time to collect some perspective. It had taken us a week to go a distance we could have driven in eight hours. We were living in schooner time now. And I liked it. Six knots across America seemed just right. I know eventually I would have become restless, I would have wanted to join the race again. I know in time I would have gotten bored. But for now the pace was comforting and natural, like a heart beating at rest.

For Hajo, it was the only pace that worked. "I'm happiest on a boat," he said. "When I'm sailing, I don't think about anything but the moment. I deal with every problem as it comes up, moment to moment. It's how I need to live. But when I've been on land for a few days, I start looking in the mirror and seeing the gray hair and wrinkles and thinking that I'm getting to be an old man. It sucks; there's too much I want to do before I die. I start going haywire. It's better when I'm at sea."

. . .

At dawn, passing Rochester, we realized that the lake was too wide to see across. At six o'clock, I carried cups of coffee below and woke Matt and Harold, a courteous touch that had been initiated by Lisa. We all appreciated her civilizing influence, though we would backslide to barbarism the moment she left the boat. I went to my cabin for a few hours sleep.

Later, I woke to a lovely morning, the lake calm, the sky streaked with cirrus. The others sat together on the deck drinking coffee and talking. I joined them there.

After eight days and 850 miles, it was time to make a decision. If we continued, we could reach the mouth of the St. Lawrence by nightfall. Then we would have to descend 744 miles of locks, canals, and river to the Gulf of St. Lawrence, followed by a couple days' journey around Nova Scotia to reach Maine. If we chose instead to take the Erie Canal, we would get off the lake in Oswego, a small city on the southeast shore of Lake Ontario, about fifty miles from the end of the lake. Hajo steered the boat toward Oswego. We would stop there and rest before deciding which route to take.

We entered the harbor at Oswego, passing the lighthouse at its mouth. Dozens of double-crested cormorants hunkered on the rocks of the breakwall, their wings held out from their bodies to dry. They seemed vaguely sinister, like rows of cloaked vultures on a parapet. Recently, a lot of people have decided that the birds are definitely sinister. In the last few decades the double-crested cormorant—one of six species of cormorant in North America, and the only one found regularly in freshwater—has gone from an imperiled species to a pest. Poisoned in the 1960s by DDT, they laid eggs with shells as thin as tissue paper that hatched, if they hatched at all, into chicks born with crossed bills and other defects. After 1972, with the banning of DDT and protection under the Migratory Bird Treaty Act, cormorants began making a comeback. But they came back too strong. From only 89 birds in all the Great Lakes in 1970, their population increased to more than 200,000 by 2001.

It would not be a problem except that cormorants are fish eaters—and they eat plenty. They also breed in colonies, and when their numbers are high, they damage trees and vegetation and attack gulls and terns trying to nest in the same areas. The U.S. Fish and Wildlife Service has allowed farmers who grow fish and other aquatic animals in aquaculture ponds to kill cormorants preying on their stocks. They've also issued permits to "oil" cormorant eggs to prevent hatching on Lake Ontario's Little Galloo Island and on an island in Lake Champlain. But commercial and recreational anglers in the Great Lakes say more severe controls are needed. They claim the birds are wiping out native stocks of perch, bass, and other fish. So far the Fish and Wildlife Service has resisted appeals to control the birds in the open waters of the Great Lakes and is busy researching the problem. Meanwhile, some anglers have taken matters into their own hands. In eastern Lake Ontario, hundreds of cormorants have been found dead of gunshot wounds, and whole colonies of nests and eggs have been destroyed.

Motoring into strong river current we could see where the rapids of the Oswego River divide the city in half, passing beneath Bridge Street in the downtown district, finally slowing at the entrance to the harbor. Docks and wharves line the river there, as do oil storage tanks, Lafarge silos, and a public promenade where couples stroll and anglers cast lures and bait—the usual awkward blend of industry and recreation. There too is a private marina, a Best Western with docking facilities, a riverside restaurant, and a Port Authority wharf where freighters load and unload cargo.

The river was high and turbulent, chocolate-colored, with half-sunk trees and logs washing down from above. Waves at midstream tossed root beer froth in the air. Small boats trying to go against the current were forced to the east bank, where the water was quieter. A channel on that side led to the bridge on Bridge Street, and the first lock of the Oswego Canal, which would be our route to the Erie Canal if we chose to go that way.

We motored to the Port Authority wharf, where the only dock space available was directly behind a large commercial tanker named *Saturn*. As we edged past, a man walked along her deck, looking down at us. He cupped his hands to his mouth and shouted: "Are you from Traverse City?" Later, when we got acquainted with him, he would tell us that he had attended the Great Lakes Maritime Academy in Traverse City and remembered the *Malabar*.

But now we had our hands full getting the boat to the dock and asked if he would catch our lines. He ran down the gangway of the *Saturn* and waited for us on the wharf. Matt twirled a small heavy ball attached to a coil of cord and flung it across the water to him. The cord was tied to the end of a dock line, which the man on shore pulled rapidly to him and hitched around a bollard. He caught a stern line and hitched it, too, around a bollard. But the current forced the schooner to veer away from the wharf, making the lines shimmy with strain. Hajo ordered Matt to run a spring line to shore and attach it to our windlass in the bow. With it we winched our bow to the dock and finally secured the boat. Hajo turned off the engine. We invited the guy from the *Saturn* to join us later for dinner.

A few people from the marina wandered over. Most of them were waiting to enter the Erie Canal. The storms had dumped more than eight inches of rain on the region and the entire canal system was closed. Rumors flew through the marina grapevine: water levels the highest in a decade; locks clogged with debris; many locks damaged and a few destroyed; a hundred boats trapped in Lake Oneida; no chance of passage for a week, at least. Meanwhile, the runoff roared down the Oswego. Locals said it was the highest they'd seen the river in years. Every log tumbling through the rapids was a reminder that the Erie Canal was unnavigable, and likely to remain that way for days.

Our dilemma remained: Wait for the canal to clear, or head for the St. Lawrence? Hajo made phone calls and confirmed that icebergs still lingered near the mouth of the St. Lawrence. None of us liked the idea of dodging ice in the middle of the night on the Atlantic. But neither did we want to sit in Oswego waiting for the water to go down while Steve Pagels called twice a day demanding to know what was taking us so long. "We'll decide tomorrow," Hajo declared. It was Saturday night. "Now it's time to dress down and libate."

That evening, Tim prepared an enormous kettle of chicken Alfredo and pasta. It fed all our crew plus a married couple from a sailboat in the marina who had come by to admire the *Malabar*, and the crewman from the *Saturn*, who turned out to be a relaxed, pleasant, quick-to-smile guy we all liked immediately. He introduced himself as Paul Garrett and said he lived in New Hampshire. He was thirty-two years old, had been working the lakes since graduating in 1988 from the Maritime Academy in Traverse City, and had recently been promoted to chief engineer of the *Saturn*. He explained that the ship was in Oswego for a week, awaiting orders for her next job. Paul had drawn the short straw. Everyone else was allowed to go home on leave.

He brought a bottle of wine for dinner and a portable CD player. We hadn't realized how starved for music we'd become. We ate on the deck listening to Mozart and agreed that it sounded better even than the Beethoven on Hajo's cell phone.

Paul told us a little about life on a lake carrier. He worked a typical schedule: four hours on and eight hours off every day for six weeks, followed by three weeks vacation. The schedule was hard on marriages. Paul's had broken up a year earlier, and he figured that the divorce rate among men on the ships was sixty or seventy percent. While on board, most of his shipmates passed their free time watching movie videos and playing computer games. He was among the minority who preferred music over television and books over computers. Not everyone could handle the life. His first year on a boat, while sailing down the middle of Lake Huron, a crewmate of his left a full cup of coffee and a lit cigarette in the galley, climbed up on deck, and never returned. Nobody saw him fall, and his body was never found.

I asked Paul about his long-term ambitions and he grew shy. Finally he admitted that he wanted to buy a sailboat and sail the world. He didn't talk about it much, he said, because most of his crewmates held private sailors in low esteem. "Rope chuckers," they call them.

Sunday we rested—did laundry, showered, walked downtown to the bookstore. Later I strolled around the marina. Oswego is a headquarters of the Lake Ontario sportfishing industry, and is crowded with anglers in late summer and autumn when Chinook and coho salmon stage offshore before running up the river to spawn. The marina is dominated by charterboats with names like *The Other Woman, Cold Steel, Lil Flipper, Hat Trick, Triple*

Deuce, Whiplash. The owners were not busy this time of year. A few hosed their boats down or sat on deck reading the Sunday newspaper. Most of the boats were empty.

Oswego began as a fur-trading post at the beginning of the eighteenth century, and was the site of the first British port on the Great Lakes. It was the terminal for travelers coming from the Hudson River up the Mohawk River, and the first major shipping port on the lakes. In 1743 it was visited by the Philadelphia naturalist John Bartram, who was making scientific observations for a book on the region's climate, soil, rivers, wildlife, and other "matters worthy of notice." In 1804, the ornithologist Alexander Wilson walked from Philadelphia to Niagara Falls by way of Oswego, sketching birds and composing lines of poetry—"Mark yon bleak hill where rolling billows break,/ Just where the River joins the spacious Lake."

For most of two centuries Oswego was among the busiest ports on the lakes. After the opening of the Welland Canal, the city's boatyards pumped out brigs, sloops, and schooners, and its mills produced tons of flour, starch, lumber, shingles, and railroad ties. Settlers passing through on their way west stocked up on dry goods made in the town's ax, scythe, cotton, and hat factories, and in its machine shops and cabinet shops and stove foundry. The Erie Canal hurt business, shuttling traffic from the Hudson River to Rochester, which became the major commercial port on Ontario's south shore, and on to Lake Erie at Buffalo. But in 1838 the Oswego Canal opened, linking the city to the Erie Canal, and business picked up until 1959, when the St. Lawrence Seaway opened the lakes to ocean traffic and Oswego was bypassed again. Today, it is a quiet port and a city of small manufacturers and tourist industries.

That afternoon, Paul led us on a tour of the *Saturn*. Though the ship is registered in Wilmington, Delaware, she has worked only in the Great Lakes. Paul gave us an insider's look at the living quarters, the wheelhouse, the enormous engines and power plant in the guts of the ship. At 384 feet she's smaller than most lake carriers, and less than half the length of many. Her usual cargo is asphalt, oil, or carbon black, which is made from petroleum residuals and is used to color printer's ink and car tires. We learned that her sister ship was the *Jupiter*, a tanker that in September 1990 caught fire and burned in the Saginaw River, a short distance from Saginaw Bay. Coincidentally, I had covered that story for a magazine in 1991. I remember being astonished to learn that petroleum-related accidents and spills were common on the Great Lakes.

The fire on the *Jupiter* had occurred only a year and a half after the *Exxon Valdez* ran aground and spilled more than eleven million gallons of crude oil into Alaska's Prince William Sound. The Alaskan accident in March 1989 focused the nation's attention on oil spills and forced American shipping companies to speed the transition to safer double-hulled tankers. It also inspired the Oil Pollution Act of 1990, which mandated the U.S. Coast Guard to establish spill response teams throughout all United States coastal regions, including the Great Lakes, where few people realized they were needed.

Unfortunately, the need was acute. According to U.S. Coast Guard documents, 3,816 spills dumped more than 1.19 million gallons of petroleum products in the Great Lakes from 1973 through 2000. Spilled petroleum in all its forms, from crude oil to fuel oil to gasoline, has more serious environmental consequences close to shore than in the open ocean and is harder to contain in freshwater than salt, putting the Great Lakes at great risk. Spills in the oceans can do enormous damage, but they ultimately disperse, whereas the Great Lakes are a nearly closed system supporting extremely vulnerable animal and plant communities and providing drinking water for dozens of cities. Some scientists have called the lakes an "ecological bathtub." An oil spill a tenth the size of the one in Prince William Sound would cover virtually all the Great Lakes and create economic and ecological devastation on a scale we can hardly imagine.

The Great Lakes sit on sizable reserves of oil and natural gas. Nobody knows how much is there, though federal geologists have estimated that 30 million to 500 million barrels of oil exist beneath Lake Michigan alone (Alaska's Arctic National Wildlife Refuge, by comparison, is believed to hold up to 16 billion barrels). Some deposits beneath the lakes have been tapped since at least 1913, especially in the Ontario waters of Lake Erie, where 550 gas wells currently extract about 10 billion cubic feet of natural gas every year, and 10 others drill for oil. Many of those Canadian gas wells are on offshore platforms; government policy insists that if they strike oil they must be immediately capped.

None of the Great Lakes states allows drilling from offshore rigs in their waters (each has sovereignty over a band extending three miles from shore), and only Michigan has allowed directional drilling from rigs set back from the water's edge. Unlike conventional wells, which drill straight into the earth, directional or slant wells burrow at angles, allowing them to reach deposits that can be miles away. As of 2002, thirteen of these rigs were in operation in Michigan, ten along the shore of Lake Michigan and three on Lake Huron. As I write this, strong public sentiment is swaying lawmakers to ban the practice altogether—or until the next energy crisis.

Most experts agree that directional drilling is relatively safe. Drilling into deposits three or four thousand feet beneath layers of shale and limestone is unlikely to cause leaks. The greater risk is at the wellheads and along the pipelines that transport the oil and gas. In spite of state-mandated precautions, including earthen dikes and impermeable liners, accidents at drill sites are a clear danger to dunes and wetlands and conceivably to the lakes themselves. Although no accidents have been reported at the thirteen shoreline rigs in Michigan, inland wells in that state were responsible for eighty-nine leaks of oil and gas in 2000, in amounts up to four hundred gallons, and for spills as large as four thousand gallons in previous years. Shoreline well operations are also ugly. They've been condemned as "sight pollution" and blamed for reducing tourism and threatening the quality of life of nearby residents.

In late 2001, when President George W. Bush signed his energy program bill, he agreed to an amendment sponsored by Michigan senator Debbie Stabenow to halt all additional Great Lakes drilling until September 2003. Already in Michigan, directional drilling had become one of the most heated environmental issues in decades. Now, for the first time, the federal government was telling the other Great Lakes states that they could not set their own natural resources policy. Critics cried that it opened the door to government control over other natural resources, including the water in the lakes. Supporters of the amendment said federal intervention is needed to protect the environment and public health.

After the publicity arising from the *Valdez* and *Jupiter* incidents, the companies that drill, pipe, and ship petroleum products in the Great Lakes made sure the public understood that more than ninety-nine percent of their spills were so-called chronic spills of less than 10,000 gallons (those greater are called "catastrophic"), and that most of them involved volumes of less than 350 gallons. But when you consider that a quart of oil makes a two-acre slick on the surface of a lake and can contaminate a quarter of a million gallons of drinking water, 350 gallons looms large.

A massive oil spill would be devastating to the lakes, but it could be dwarfed in significance by a spill of toxic chemicals. Ships, trains, and pipelines transport millions of gallons of chemicals through the Great Lakes basin. Because many of the chemicals are invisible and easily dissolved, they are considerably more difficult to contain and clean up than oil. A chemical spill near a city's water-intake pipes could bring the city to its knees. And if the chemical were highly toxic, its impact might be felt for decades.

But it was gasoline that caused the problems on the *Jupiter* that day in 1990. The ship was moored at a dock in the Saginaw River and had unloaded about half her cargo of 2.3 million gallons of gasoline when a fire broke out. Analysts still disagree about how it happened. The captain and crew claimed the wake from a passing freighter tore the *Jupiter* from her moorings. The captain of the passing freighter, the *Buffalo* (a vessel that would make news again a few years later when it ran over a Detroit River lighthouse), insisted that he was traveling upriver at barely more than an idle and could not have produced a wake large enough to rip the *Jupiter* from the dock.

Whatever the cause, the fuel line from the ship to the storage tanks on shore snapped and gasoline spewed over the dock. Sparks from an electrical connection ignited it, and the flames jumped to the ship. The fire raged for thirty hours. Firefighters finally managed to put it out, but it reignited and burned for another two days. Six to seven thousand gallons of gasoline escaped into the Saginaw River, where most of it was contained. The ship was destroyed.

All this I remembered, based on a few interviews and photographs. But if my memories of the incident were clear, Paul's were incandescent: He was aboard the *Jupiter* that day. He is convinced the *Buffalo* was going too fast up the river and passed too close. "It sucked us right off the dock," he said. "Pilings broke, the fuel line parted, and sparks got to the gas. Next thing I know, alarms are going off and people are yelling that there's a fire. Believe me, that's the last thing you want to hear when you're on a boat carrying a million gallons of gasoline. If there had been less gasoline and more fumes in the hold, the explosion would have killed us all. As it was, fire spread quickly through the boat. I had just enough time to put on a life jacket and run to the stern with a bunch of my crewmates and jump. Some of them were hurt in the jump. We didn't realize it until later, but one guy drowned."

Lisa's father, Walt, arrived in Oswego that afternoon and drove us to a shopping center across the city to stock up on groceries and personal items. Later we ate dinner at Coleman's Irish Pub, in a tall brick building built in 1828 at the edge of the river. Over drinks, Hajo called for a vote to decide whether to wait for the Erie Canal to clear or push on to the St. Lawrence. After brief discussion, a show of hands made it unanimous: We would take the Erie.

I had a personal decision to make as well. Originally, I had planned to ride the *Malabar* only to the end of the Great Lakes. Continuing on through the Erie Canal, the Hudson River, and the Atlantic was unnecessary and probably frivolous. I had a book to write, a family to support. How could I justify three more weeks away from home?

But I couldn't leave the boat now. In a day or two we had to take the rigging down, and the job would require all the help Hajo could get; it would be heartless of me to duck it. I had already suggested to my wife on the phone that I might have to go as far as the Atlantic, to see how the story of the *Malabar* unfolded and to get some saltwater perspective. She laughed at my clumsy rationalizations and said I would be a fool not to finish the journey.

Now, as if reading my thoughts, Hajo took me aside and said that Harold was planning to get off the boat in New Haven, and Tim was talking about joining him. If I left, Hajo and Matt would have to sail the rest of the way to Maine alone, which would be difficult at best. Hajo had talked to Steve Pagels about the problem and Steve offered to pay for my flight home if I stayed with the boat all the way to Bar Harbor. As a bonus, he would buy me the best lobster dinner of my life. I couldn't refuse.

We gathered on the sidewalk outside the restaurant and said our good-byes to Lisa. She'd made all our jobs easier. We would miss her skill and cheerful attitude. Hajo was convinced she had brought good luck. She hated to leave, but it was time to get back to work. Children in Maryland needed her.

That evening on the *Malabar* I learned I was not alone in thinking Hajo's muttonchops and blue eyes made him look like a classic sea captain. A few years earlier, he'd been cast as an extra in Steven Spielberg's *Amistad*. He was also hired to sail a pirate ship destined to be sunk in one of the "Hornblower" episodes on television's A&E channel. He took that job not for glory or money, he said, but because he admired the boat and wanted to sail her.

Matt said he too had movie experience. He had appeared in a made-for-television Italian movie—a spaghetti pirate film—and in a Japanese movie about the first European ship to sail to Japan. He had also crewed aboard the *Golden Hind* in several films, including the "Hornblower" series on A&E.

Then Harold said he'd also been in a movie. As a hobby he is a Civil War reenactor, one of those guys who on the weekends dresses up in Union blue or Confederate gray and reconstructs famous battles. Harold admitted he was sort of a "Farb," a name zealous reenactors give those who are in it merely for kicks. But he played the role well enough to earn five dollars a day as an extra in the film *Gettysburg*, playing the role of cannoneer for both the Confederate and Union armies. His proudest moment came during the filming of Pickett's Charge, when he ran past in the background, grinning at the camera. If you look closely, you can see him.

By Monday morning, Memorial Day, the river had lowered considerably. The standing waves in the rapids were smaller, and a buoy that had been tipped and churning in the strong current now rode upright. The current still carried debris down from above, but at least we weren't seeing whole trees go by. Everyone in the marina was confident we would be allowed to enter the canal by Wednesday.

The Erie Canal is a shortcut, though it comes at a price. Low bridges along the route mean that sailboats have to lower their masts before entering the canal system. For a small boat, that's a relatively easy job, but for one the size of the *Malabar* it can be extremely difficult. Hajo had already spent hours on the phone talking to Port Authority officials about using their crane to lift the two masts from the boat. To Hajo's delight, the director of the Port Authority offered to do the job gratis, in the spirit of promoting tall ships on the Great Lakes. His crew would be ready for us Tuesday afternoon. That gave us plenty of time to take down the rigging, we thought.

Every step in the sequence of tasks was choreographed by Matt. First we removed the foresail and the jib club and all the rigging on the foremast (we had never raised the topmasts; they lay strapped to the deck the whole trip). We stowed the club and boom on deck, as much out of the way as possible, then coiled the rigging and folded the sails and stored them with the halyards and blocks in the forecastle. We had hoped to get to the mainsail rigging also, but ran out of daylight.

For dinner, Tim grilled burgers on the deck. Paul joined us again, and again brought his CD collection. We stayed up late talking about boats, books, and weather, that great triad of sailor's topics, with Van Morrison singing in the background. At one point an argument erupted over who was the better author, Patrick O'Brien or C. S. Forrester. The battle was fought with daggers and belaying pins, but the captain's saber drove home the deciding point, in favor of O'Brien.

Tuesday, we worked twelve hours without break to downrig the mainsail and step the masts. It was a harder job than we'd anticipated. Hajo yielded to Matt's greater experience, and stood by to assist when needed. Matt was magnificent and obnoxious. He swaggered, he bragged, he yelled at Harold and Tim, he demanded constant recognition for the job he was doing, and he did it methodically, safely, and brilliantly.

Hoping to replace Lisa, Hajo had made some calls and talked an old friend into joining us for the next leg of the trip. Tom Hennessy, from Colts

Neck, New Jersey, showed up at the wharf just as the crane was getting ready to lift the first mast from the boat. He was middle-aged and professorial, dressed in a button-down shirt and chinos, but not afraid to get dirty. He rolled up his sleeves and went to work.

The mast was stuck so tightly that the crane had to jerk it free. The sudden release of weight made the *Malabar* surge and rock at her moorings. That was hard on Hajo. His emotional attachment to the *Malabar* had become powerful. As the mast was yanked free and lowered to the wharf, he grew so agitated he couldn't watch. He kept walking away. Finally I asked what was wrong.

"I hate this," he said.

"Hate what?"

"This. What we're doing to her. We're ripping her apart. It's like a rape. She's being violated."

By evening, both masts were lashed tight to the cabin tops. They reached most of the way from the poop deck to the bow.

Nobody had the strength to tell stories for dinner that night. We ate Tim's "blackened Cajun" meatloaf—burned when he had to leave the oven unattended while helping on deck. It wasn't quite dark yet when we collapsed into bed.

Early Wednesday morning we walked along the river, talking with people. Several owners of private boats were getting ready to enter the locks and head up the canal. The rumor was that those boats that had been waiting longest would be allowed to go first. We would be in the first group. One waiting powerboat was owned by a stocky man with a Maine accent who turned out to be a lobsterman. He and Hajo discussed the differences between lobstering in Maine and Long Island Sound. Hajo asked him what he thought of the Great Lakes and the man went silent for a moment. He had just purchased his boat in Rochester and had motored it along the shore to Oswego.

"It's strange," he said, shaking his head in bafflement. "There's waves but no swells. And there's no *tide*."

Then it was time to go. We climbed aboard our suddenly streamlined schooner, the deck so cluttered with masts and booms and fuel drums that getting to the cabin passageways or from one side of the boat to the other was like scrambling over logjams. We would have to live with it until Albany.

Paul Garrett cast off our dock lines and walked along the wharf beside us. He was owed a week's vacation, he said, and he would try to join us on the Hudson. We hoped he could.

We crept into the first lock in company with two small sailboats, were lifted eight feet, and entered the Oswego Canal. Beside us and at a lower elevation were the rapids and waterfalls of the Oswego. Eight locks upstream the Oswego Canal became the Oswego River, wide and meandering, deep on the outside of every bend and shallow on the inside. Twenty-four miles above the city of Oswego we came to a junction with another waterway and turned east. It was an unremarkable, currentless, tree-lined channel filled with debris and muddy water. The Erie Canal.

Chapter 14

ERIE CANAL/HUDSON RIVER

The Erie Canal and the American Dream ◆ We Run Aground Again ◆
Raising the Sticks in Albany ◆ Down the Hudson at Night ◆ A Manhattan
Morning ◆ Hell Gate to the Sound

All that first day on the Erie Canal, Harold and I leaned from the bow
with poles and nudged logs and broken trees aside. The canal, running full
of dark water, seemed more a natural river than an artificial one. It mean-
dered through woods, its shores canopied, its banks dense with vegetation.
Nature had won again: the labor of thousands was forgotten. Bankside
towpaths once tramped by mules and horses had grown over with jungles
of shrubs and vines as lavish as those that swallowed Mayan cities. We
entered a mossy concrete lock with iron gates pitted with age and were
raised seven feet to the next level. As the water poured into the lock chamber, we watched a man in a front-end loader claw driftwood from the water.

We had entered the canal at about its midpoint. If we had chosen to
enter from Lake Erie, at the mouth of the Niagara River near Buffalo, we
could have bypassed the Welland Canal and all of Lake Ontario and followed the canal its full 360-mile length to the Hudson. But travel on the
canal is prohibited at night, so we saved time by entering the canal from
Oswego.

Though the Erie Canal is used primarily now by pleasure boaters, it was
built for commerce, that cold science ruled by Basic Transport Units (*x* tons
moved *y* miles in one man-hour). From its conception the canal was cham-
pioned tirelessly by New York statesman DeWitt Clinton, but he was up
against opposition that delayed construction for years. Thomas Jefferson

denounced the canal as madness, and the press vilified it as "Clinton's Big Ditch" and "Clinton's Folly." When ground was finally broken in 1817 in the town of Rome, it was on flat terrain calculated to get the enterprise off to an easy start.

At a time when it cost less to ship cargo across the Atlantic than to send it by wagon a couple dozen miles inland, the Erie Canal made it affordable to transport goods and passengers to the interior of the continent. Before the canal, sending a ton of cargo by horse and wagon from New York to Chicago cost about thirty dollars. That same ton cost a single dollar via the Erie Canal. America would never be the same. The project made New York the Empire State, transformed New York City's harbor into the busiest in North America, and galvanized the nation.

The work of digging the canal and building its locks was accomplished by a force of about nine thousand laborers, many of them recent immigrants, each earning wages of ten to fourteen dollars a month, plus food and lodging and half a pint of whiskey a day. The waterway was soon recognized around the world as a major technological achievement, the most significant in the young nation's history. "American ingenuity" became a catchword of the day. Engineers with no formal training in engineering surveyed the route, designed locks and aqueducts, invented machinery and tools as the need arose. People were amazed not only at the scale of the canal but at its apparent defiance of natural laws. In an era when nature was usually perceived as an enemy to be vanquished, the Erie Canal had triumphed. It forced rivers to run uphill and carried them across valleys; it allowed boats to travel across country that had been wilderness only a few years before.

Workers dug the canal to four feet deep and forty feet wide (by 1835 it would be enlarged to seven feet deep and seventy feet wide). But that was the easy part. A 568-foot rise in elevation from the Hudson River to Lake Erie, and the many incidental valleys and hills along the route, required eighty-three locks be built. In addition, eighteen aqueducts were needed to lift the canal as high as thirty feet and carry it up to a thousand yards across valleys that bisected the route. At Lockport, two sets of five locks each were built, one following another, raising and lowering boats sixty feet. Three miles of "deep cut" there had to be blasted through solid rock.

When the canal was completed on October 26, 1825, signal cannons placed every few miles along the route sent a series of blasts from Buffalo to Albany and down the Hudson to Manhattan, then back again to Buffalo—an audio relay that rumbled a thousand miles in less than three hours. While the cannons boomed, DeWitt Clinton and other dignitaries set off from Buffalo in a packet boat, the *Seneca Chief,* followed close behind by a tugboat decorated as Noah's Ark and filled with birds, mammals, "creeping things," and two Indian boys decked out in tribal dress. Thousands of people gathered along the route to watch and cheer.

In New York, amid a frenzy of fireworks, Clinton unstopped a cask of fresh water from Lake Erie and poured it into New York Harbor. His friend and fellow canal promoter, the scientist and politician Samuel Latham Mitchell, emptied into the harbor bottles of water collected from the great rivers of the world—the Nile, Ganges, Indus, Gambia, Thames, Rhine, Seine, Danube, Orinoco, La Plata, Amazon, Mississippi, and Columbia—and declared "the circumfluent ocean republicanized. It is done. I pronounce this union blessed." Even before the celebration ended, the first boatloads of goods were being prepared for shipment to Buffalo.

The canal created strong local economies. Many of the towns along its course grew up during and soon after the construction, especially in the vicinity of the locks, where boat crews and passengers would purchase food, souvenirs, liquor, and other goods while waiting for their barges to negotiate the locks. Any towns and villages with the suffix "port" in their names were probably established around this period. Lockport, Brockport, Weedsport, Spencerport, Middleport, and Gasport all began with the hope that they could capitalize on the bonanza. And the bonanza was real. Lockport grew from a cluster of three families in 1821 to a boomtown of over three thousand people four years later. By the middle of the nineteenth century, the corridor along the Erie Canal was inhabited by hundreds of thousands.

In 1825, before the canal was even in full operation, more than forty thousand passengers rode it. In the decades that followed, millions of immigrants, settlers, entrepreneurs, and merchants piled onto barges and made their way to and from the Great Lakes. The canal was a final leg of the Underground Railroad, carrying runaway slaves from Syracuse to Buffalo, where they crossed to safety in Ontario. Soon it became the central feature of a popular "northern tour" that climaxed with a visit to Niagara Falls.

It was quick and easy transport for the times. Packet boats measured up to seventy-eight feet long and fourteen wide and could accommodate 120 passengers during the day and sleep 40 at night. Towed by teams of horses on towpaths, they reached speeds of five miles an hour. Freight boats were heavier and made about two miles an hour with their loads of wheat, oats, logs, and commercial goods. The 363-mile trip from the Hudson River to Buffalo could now be made in five to seven days, twice as fast as wagons. Canal boats ran day and night and departed at intervals of a few hours, whereas stagecoaches and wagons stopped for the night and could be boarded only once or twice a day.

For immigrants seeking opportunities in the cheap lands to the west, the canal was a godsend. But what some saw as a road to opportunity, others saw only as a ditchful of unsavory characters and filthy water. Nathaniel Hawthorne traveled the canal in 1830 and complained that it was unpleasant, that it had given birth to poorly developed towns and had destroyed the natural beauty of the landscape. Passengers on packet boats were often disillusioned. Crowded boats and a shortage of beds required many travelers to lie on the floor at night, and even those lucky enough to secure a bed had trouble sleeping because of the noise, heat, stagnant air, the mosquitoes. Low bridges were a frequent hazard. Farmers separated from their fields by the canal threw up hasty bridges and built them not an inch higher than necessary. Passengers who climbed to the tops of the packet boats seeking fresh air were sometimes crushed or swept overboard.

In four days on the Oswego and Erie canals we caught a glimpse of America as it was seen by countless immigrants who followed this route to the nation's heart. They journeyed from Ellis Island up the Hudson to the Erie Canal to Lake Erie and across the big lakes to the forests and prairies of the Midwest and beyond. My Irish ancestors probably reached Michigan this way. My wife's Swedish great-grandparents likely rode a wave of immigrants recruited from Scandinavia to work in the lumber mills along the Manistee River in northern Michigan. They almost certainly arrived there via the Erie Canal, with their families and all their possessions loaded on canal boats, then transferred to steamers or schooners in Buffalo. I could imagine their excitement and fear and optimism. How vast the continent must have seemed—and how vast the possibilities.

Something of the vitality and optimism of a young nation hangs over the canal still. It was among the first of a century's worth of revolutions in transportation and communication. It was soon followed by rails, then by telegraph wires, roads, power lines, telephone lines, highways—lines of various kinds reaching across the continent and ultimately connecting other continents until the planet was networked into a smaller place. The optimism was both practical and idealistic. Commerce came first, but lines of connection and transport made it easier to exchange not only goods but ideas, for the general betterment of the world. Settlers scattered across spacious young America must have been grateful for every canal, telegraph wire, train, and road—any innovation that might make their lives less difficult and less lonely.

As the Erie Canal nears Oneida Lake, it grows wider and is bordered with houses and cottages. Hajo, standing at the helm, commented on them. "These houses are all perfect examples of people finding the American Dream," he said. "A place to get away, a place on the water. Little houses, some of them modest, some show-offy, each with a lawn and a dock. For me, I want the perfect boat. I don't want to be encumbered with land. I want a boat perfect for my needs, then I want enough money to do this"—he waved his hands to include the *Malabar,* the crew, the Erie Canal—"for a living."

"But you *are* doing this for a living," I said.

"I know. I mean a *better* living."

We crossed Oneida Lake, following navigation buoys down its middle to avoid shoals. The cool breezes of the big lakes were behind us now, replaced by heat and sun. Matt tanned even darker than before, while the rest of us burned red and peeled. Taking bow watch required slathering all exposed skin with sunscreen and wearing a hat and sunglasses. Matt jury-rigged a canvas Bimini cover over the poop deck, and we gathered in its shade and drank iced tea.

As always, Hajo set the mood. The rest of us adjusted to him and were pleased when we received his approval. Like all natural leaders, he earned respect effortlessly. His strengths were quiet strengths. He did not judge. He was not willful. When it came to delegating jobs, he focused on our individual strengths and overlooked our weaknesses. If we did our jobs well, then we were good for the boat, and that, in Hajo's world, was among the highest accomplishments. A good crew was as rare as a good boat. But you don't discard a boat because it happens to have flaws. Minor flaws can be fixed or ignored. Likewise with people.

He had his breaking points, however. One morning I reported that the aft midship head was not working. Hajo's response was explosive.

"Someone is not using it properly! There is nothing wrong with that head!"

It turned out I had been using it improperly all along. I'd been holding the pump pedal down while I operated the flusher, causing water to surge into the bowl at the same rate it was being pumped away.

But there was another problem. Tim had been complaining for a week that the stench of the head was getting worse, that the holding tank must be leaking. The rest of us were beginning to notice, too. Hajo refused to listen. Long after it was impossible to ignore the problem, he persisted.

"We'll check it out in Albany," he said.

Near Rome, at 420 feet above sea level, we reached the "hump" of the Erie Canal. I'd been curious about it. Water at the top of the canal system should run off in both directions. What kept it deep enough for navigation?

The answer was simple. At the point of highest elevation the Mohawk River enters from the north, supplying water that is diverted west to the Great Lakes and east to the Hudson. We crossed the top of the watershed—all water from now on flowed to the Atlantic—but the current was so sluggish we didn't notice the change. Instead, we watched for old barges abandoned in frog-water off the main canal, some of their skeletal remains so old they sprouted trees a foot in diameter. A few stretches of towpaths here had been cleared to make bicycle and hiking trails.

Every lock from now on stepped downward, each of them lowering us sixteen or twenty feet toward the Hudson. The water remained full of debris, though tons of it had already been removed and left in heaps near each lock. Banks had washed out and been hastily repaired. Clumps of debris hung in trees ten feet above the water.

The Erie Canal was not designed for large boats. The *Malabar*'s length of 115 feet fit the locks with dozens of feet to spare, but drawing 8.5 feet of water could be a problem at normal water levels. It was a problem even in high water.

We were passing marker 505, somewhere between Locks Nineteen and Eighteen, when we saw two large motor yachts approaching from ahead. All day we'd been meeting other boats. It was fun to check out their ports of call. One cabin cruiser was from Charlevoix, Michigan, just a few miles up the shore from Traverse City. The people on board recognized the *Malabar* and shouted to us in greeting.

The two boats now approaching appeared to be in the forty-foot class—new luxury models with large cabins and flying bridges. They barreled upstream much faster than the six-knot speed limit. Large wakes fanned behind them and washed over the banks into the woods, making small trees sway. The second vessel slowed as it approached us, as it should, but the first maintained its speed and refused to yield the center of the canal. Hajo had no choice but to steer to the right, farther than he wanted to go, so far he was shouting in fury as the yacht roared past, its wake shoving us even farther aside.

Our luck turned bad. At the spot where we were forced closest to shore, a tributary creek entered the canal. At its mouth was a fantail of sand and gravel. Our depth locator read ten feet, nine feet, eight, seven. With stomach-dropping finality we slumped to a stop.

Hajo was livid. "Somebody get the name of that boat!"

But it was gone too quickly. It continued on without slowing, without anyone on deck even glancing our way. The second boat, the one that had slowed, continued on also, the man at the helm looking back as he accelerated.

We lowered the yawlboat and tried to push the larger boat, but it would not budge. We stretched a two-inch line from the windlass in the bow to the nearest stout tree on the bank. Hajo reversed the engine as hard as he dared, while Tim, Matt, Tom Hennessy, and I cranked the windlass. The line stretched so tightly that a fine spray of water shimmered around it. Still the boat wouldn't move.

Our radio wasn't strong enough to reach help, but a private sailboat hailing from Sodus Point, New York, came along and relayed messages to the canal authorities and told them our location. Soon a tugboat was on its way.

Tender 2 was so compact and colorful that it could have been the star of a children's book. But it was a workhorse. With its crew of three, it tugged mightily and pulled us off the sandbar. We shouted over the engine noises to ask the names of the crew, but all we could hear was the skipper saying he was Mike something, a Polish name a yard long. The lockmaster at the next lock told us everybody on the job knew him as "Mike with a Diddle in the Middle."

As dusk fell we approached Little Falls, New York, a mill town on the Mohawk River. We drew up to the public wharf. A dozen local residents sat on lawn chairs, fishing, their rods propped against forked sticks, kids running around, men drinking beer and watching the river, women chatting among themselves. We hated to break up the gathering, but we had no choice. The wharf was the only dock space large enough to accommodate the *Malabar*. We drifted past, turned around in midriver, and came up against the current. As we got close, the people on shore stood up, reeled in their lines, and moved their chairs back from the edge. Several of the men stepped up to catch our dock lines.

When we were secure, a small crowd came forward to look at the boat. Sailors from other boats came for the usual banter. We told them about running aground, and they said that they'd been watching motor yachts speed past all day, apparently trying to make up the time they lost while the canal was closed.

We mentioned that we needed groceries and showers, and somebody pointed to the community building in the park, which contained a single shower, and somebody else said that the nearest grocery store was a long walk down the road, across a bridge, and downtown.

A young man named Dan stepped forward, his fiancée Gloria in shy tow, and offered us a ride. Matt and I hopped in the back of their pickup and Gloria's ten-year-old daughter Breanne drew open the little window at the back of the cab and chatted with us as Dan drove across the bridge to the downtown district. Breanne pointed to a mansion on a hill surveying the town and said, "That's where the Millionaire lives." "Who's the Millionaire?" I asked, and she said she didn't know his name. "How do you know he's a millionaire?" I asked, and she said, "Look at the house, silly." She grinned and asked what it was like to sail on a pirate ship. Dan said she shouldn't ask so many questions, and Matt and I said it's okay, it's great, and told her about the boat and promised to give her a tour when we got back. I asked Dan what he did for a living. He worked nights in the pulpmill and had been putting in a lot of overtime lately, which he needed because of the wedding coming up and he and Gloria hoping to buy their own house. Gloria sat in the passenger seat smiling, but wouldn't say a word.

At the supermarket, Matt and I careened up and down the aisles and heaped meats, vegetables, bread, and beer in a cart. As a gift to Dan, Gloria, and Breanne, we bought fresh strawberries, whipping cream, and shortcake. We returned to the boat and gave Breanne and her family a tour. Finally, Matt and I ate a quick reheated dinner on deck and took turns getting showers at the community building.

In the morning we were awakened by a rattling good thunderstorm. We waited it out, drinking coffee in the main cabin. When the rain tapered off we threw off our lines and got under way, passing beneath cliffs and lush, dripping forests, the Millionaire's mansion looking down on us from the bluffs above the city.

We saw Little Falls themselves, a series of nearly dry precipices, most of the water that once roared over them diverted into the canal. Originally an immense stone aqueduct had shunted the canal over the falls, but in 1917 it was replaced by a forty-foot-deep lock that was at the time the largest in the world.

Below Little Falls the Mohawk River is wide and slow, with buoys marking the navigable depth at midstream, and locks and dams spaced at frequent intervals. The river here is paralleled closely by busy Highway 90, the

New York State Thruway. On the other bank are train tracks. We kicked up the engine a notch and cruised, making good time.

With binoculars we could read traffic signs on the Thruway: 42 miles to Albany, 190 to New York City. A few hours by car or train, but it would take us nearly a week.

Running aground had set us back half a day, but we had our legs beneath us again and were feeling good. We knew the vessel and what she could handle and were certain we could solve any problems we encountered. Six knots across America was no longer a lark—it was a valid means of transport. Only a few people seemed to know it. The rest must have forgotten. Most of us long ago accepted rapid transit as a way of life, and accepted without question the price we have to pay for it. Flying to Disney World for a week or carpooling from home to work every morning are often antagonistic and gut-wrenching ordeals, exhausting, depleting, rage-inducing, treacherous—and we seem to have no other choice.

But the *Malabar* had shown us a better way. It was as if we'd discovered a secret passage behind the plaster-and-lathe hallways of the world. No longer did we have to endure the overcrowding and chaos of mainstream routes. No airport delays, no lines at the ticket counter, no bumper-to-bumper commuter mayhem of vehicles driven with aggressive mastery by grim and merciless people, and God help you if you try to change lanes in front of them.

We were free of all that. We would return to it soon enough, of course, but for now we'd found an alternative. We sat back and watched cars and trucks stream past a few hundred yards away on Highway 90. On the other shore was the railroad. High above us jets drew contrails across the sky. In the midst, at the center of it all, we had found calm.

Then something astonishing happened. On the highway, a truck passing at ten times our speed gave three long blasts of its airhorn. The driver raised his arm from his window in a salute. A few minutes later another trucker did the same—then another and another. They blew staccato or blared continuously or performed baritone melodies, shave-and-a-haircut-two-bits. On the opposite shore an Amtrak train barreled past and the engineer gave ten long blasts of the horn. He put his arm out the window of the engine and raised his thumb.

Hajo danced a jig on the poop deck. Matt spun to wave at the trucks on the highway, then at the train, then at the trucks again. Harold tipped his head back and laughed. Tim and Tom Hennessy shook their heads, disbelieving.

We'd become the star exhibit in the history of mechanical transportation. From here to Manhattan, whenever our route ran beside highways and railways, we were saluted. A few days later, in a restaurant, I would get into conversation with a truck driver and ask him why he thought truckers kept honking and waving at us. He would say, "Because we admire your freedom. Because it's getting too damned hard to make a living driving truck. I'd love to do what you're doing. Sailing free and easy. We envy you guys."

Rain started falling again, a downpour just as we entered Lock Thirteen, but not even that could dampen our spirits. We pulled beneath a bridge and waited it out laughing, raingear drawn over our heads, streams of gutter-flow pouring from the highway above and plunging into the river. Tim came up from below with cups and a carafe of coffee, and we drank it while the rain poured down and the debris in the river piled against the coffer dam. Below the dam the river flowed high and yellow, with big tufts of foam riding the current like couch stuffing.

We descended through the lock and motored past the unhealthy waterfront of Amsterdam, a once-thriving industrial city. Its factories are abandoned now, their windows broken, their parking lots empty and choked with weeds. Freight trains rumbled past without stopping, reminding the city that industrial life goes on, but elsewhere.

The river, we noticed, was dropping. It dropped so far that mudflats gleamed along the shores and old washing machines and car tires poked out of the shallows. Only in the center of the channel was there enough water to carry us. We later learned that debris had jammed open the gates of the control dam at the next lock downstream and it was allowing the river to pour through. Now we eased along at half speed, Hajo and Matt monitoring the depth finder, the rest of us in the bow with poles, pushing logs aside. Some kids ran along the shore, splashing to their knees in mud, and threw rocks that fell far short of us.

That evening we tied up to a wharf at Lock Eight, just above Schenectady. Tim served Greek salad, deep-fried walleye, and slabs of tomato sprinkled with parmesan and fresh basil. We ate on deck, swatting mosquitoes while the sun went down.

In the morning, we reached the Hudson. The Mohawk enters it without ceremony, a simple merging of rivers. A sign on a point of land points north to Lake Champlain and west to the Erie Canal. The Hudson was smaller than I expected.

Just below the junction is Troy, its old brick factories abutting the river, then two miles downstream half the Hudson tumbles in foam and roar over a waterfall and the other half funnels into the last lock in the system—the first, if you're going upstream. Federal Lock Number One should have been a cakewalk, but it was the trickiest lock of the trip. Because the Hudson was high, its current caught us midship and sucked us into the wharf. Our bumpers rubbed the wall, wood and concrete howling where they met. It ripped our fender boards to pieces. The feedbags of straw exploded into puffs of chaff that swirled in the wind. The lockmaster ran toward us waving his arms. "Get off the wall! Get off the wall!" he shouted, as if we were a twenty-footer that could be pushed by hand. We laughed at the man. His indignation made him foolish. We ground to a slow halt and threw our lines over cleats on the wall. Straw drifted in the air.

"Don't tie to the handrails," the lockmaster said and walked away. We tied to the handrails anyway.

Late in the afternoon, we pulled up to an empty Port Authority wharf in the industrial heart of Albany. Across the river, a quarter of a mile away, was a junkyard where an electromagnetic crane swung wrecked cars into a crushing machine.

The river itself was crowded with speedboats and motor yachts and 500-foot barges pushed by tugs. We had no choice but to tie the *Malabar* broadside to the river, so that the wake of every passing boat set us banging against the wooden pilings of the wharf. Hajo had called ahead for permission to dock here and to reserve a crane and a crew of stevedores to raise the masts on Monday. This was Saturday, so we had a couple days to wait.

Even 150 miles up the Hudson, we could feel the ocean's pull. The tide in Albany rises and falls four feet, twice a day. It was low tide now, and the wharf stood so high above us that Matt had to shimmy up a piling to get there. We slung dock lines to him and he hitched them to bollards the size of fire hydrants. Matt suspended a hanging ladder and we climbed up.

The wharf was enormous, and appeared deserted. Spaced along it were four warehouses, each as big as a city block. We stepped inside the nearest one. It was empty except for a stack of raw cocoa beans in burlap sacks. They made a pile the size of a house, but it barely filled one corner of the warehouse. A dozen schooners could have fit in there. Outside were vast expanses of asphalt, chain-link fences in the distance. Beyond the fences were the domed tops of petroleum storage tanks. Somewhere beyond that was the city.

A watchman wandered over to check us out. He had hair to his shoulders, tattoos on his neck and arms, six silver earrings along the rims of his ears. He became friendlier after we explained why we were there. He offered us the use of the telephone in the gatehouse, an aluminum-and-glass cubicle where he passed his shift every day watching television and talking to his friends on the phone. He warned us about the neighborhood beyond the fence, said people out there would kill us "for the gold in our teeth." He gave us a key to the restroom in the cocoa warehouse and offered to supply us with marijuana and hookers, if we needed them. We said we'd keep it in mind.

After dinner we said good-bye to Tom Hennessy, who'd been good company on the Erie Canal. A cab pulled up on the wharf and took him away.

An hour later, another cab drew up. Out stepped Paul Garrett, our friend from the *Saturn* in Oswego. He'd come to spend his week's vacation with us, as he'd promised. He tossed his suitcase to us on the boat and climbed down the boarding ladder.

Paul stopped short when he went below deck to claim a cabin. The midship head was by now officially broken—we were using only the forward head, and only in emergencies—and the stench around it was overpowering. Matt, Harold, and I had become inured enough to climb into our sleeping bags at night, cover our heads, and sleep. But it was too much for Tim, whose cabin was nearest the head. After dinner he climbed on deck carrying an overnight bag and said he could not tolerate life in a sewer. He warned Hajo that unless the head was soon fixed, he would leave the boat for good. He called a taxi and left to spend the night at a hotel.

We worked on the problem that evening, but made little progress. The holding tank was so full it was leaking, but we had no facilities to empty it. Until we found a marina with a pumping station, our only choice was to live with the stench. Out of consideration for Tim we tightened the seal at the top of the holding tank, mopped the floor, and sprayed air freshener around.

The next day, the temperature fell into the low fifties and rain returned. It would not stop for three days. The river was already high and filthy— mud brown and filled with debris and with dead and bloated shad and carp. The waves from boat traffic washed beneath us. At night, trying to sleep, we rolled back and forth in our berths and had to brace our arms against the walls to keep from falling out. We woke every time the boat crashed against the bumpers on the pilings. Hajo's and my cabins were nearly rain-proof. The roofs of the others leaked, soaking their sleeping bags and cloth-ing. The hours began to grow long.

Monday, we raised the masts. The stevedores and crane operator worked well with us, listening to our concerns, doing their jobs with care. Matt and I crawled into the hold to align the butts of the masts as they were lowered one at a time by the gigantic crane. A half-dozen dockworkers on deck yelled instructions to the operator while Matt and I shouted up requests to twist the mast a quarter-turn clockwise or move it forward two inches.

Tuesday, we hoisted the headgear to the tops of the masts, set up blocks, adjusted the shrouds. The work was tedious, made more difficult by the rain. It was one of those rains that penetrates even the best raingear, soaking you first along your arms when you raise them above your head, then soak-ing through the seams at the shoulders, neck, back, and thighs. We worked wet, cold, and miserable.

At some point we discovered the river was full of condoms. There were dozens of them, hundreds, all swimming languidly downstream, like latex jellyfish. Matt called them "Hudson River trout." He said, "Hey, condom season is open," and "Look! It's a condom race." The rain was overwhelming the city's storm sewers.

We were seeing the river at its worst. The Hudson has been cleaned significantly in recent years, but like rivers and harbors in the Great Lakes watershed, its bottom sediments remain contaminated with toxic chemicals. Upstream from Albany, between Troy and Fort Edward, forty miles of the Hudson were polluted with PCBs that a General Electric plant discharged into the river from the mid-1940s to the late 1970s. In 2001, the EPA announced a plan endorsed by Governor George Pataki to make GE responsible for dredging about 1.3 million pounds of PCBs from the river's bottom and disposing of it on land, a cleanup that could cost the company as much as $490 million. Feelings were running strong both for and against the planned cleanup. Environmentalists applauded it, but local advocacy groups protested that dredging would disrupt business along the river and stir up sediments that might better be left untouched.

That night I woke to the sound of rain hammering against my cabin roof. The river sounded fuller, perhaps in flood. I wondered if the rising water had caused a dam somewhere upstream to burst. The more I thought of it, the more likely it seemed. I imagined a wall of water rushing downriver, striking the *Malabar* and capsizing her. In the distance came a rumbling, droning noise. It sounded like a tug barge out of control in the heavy current and bearing down on us, and it seemed to be getting closer. Suddenly, something on deck broke with a splintering of wood. I jumped out of bed in my longjohns and scrambled up the ladder.

Everything was fine. The river was lit with the dim yellow glow of city lights and the rain fell steadily. The water had risen to its highest level since our arrival—the deck of the *Malabar* was now flush with the top of the wharf—but no vessels approached, no waves of floodwater rushed toward us. The engine sounds I'd heard were the normal sounds of a port city at night. The splintering wood had been the boarding ladder. As the river rose, the ladder went slack and doubled up between the boat and a piling;

a rung got crushed when the boat shifted. I went back to my cabin and slept.

We left the next day in sunshine, casting off into the swollen and dirty Hudson and steering downstream. Hajo, Matt, Harold, Tim, and Paul—all of us were excited to be under way. A few miles downstream we swung around and came up against the current to a marina, where Hajo performed the toughest dockage of the trip so far, against a low floating dock half the length of the *Malabar*, squeezing between pleasure boats moored at other docks close to the bow and stern. But *Malabar* came against the dock with a gentle kiss and stayed tight, despite the surging current. It was like watching an airplane swoop down from the sky and land on a pontoon boat. Hajo said again that he did not control the boat: "She does what she wants to do. She tells me if she wants to dock. If she doesn't, there's no way I can force her."

We took on diesel, gasoline, and fresh water. The marina had pumping facilities, so we uncoiled a three-inch hose and inserted it into the holding tank and turned on the pump. But all we managed to do was compound the mysteries of our cantankerous waste system. No matter how much foul water we removed, the holding tank remained full. Eventually we would discover that water was leaking into the tank through an intake line under the hull, but we had no way of knowing that now, and no way to fix it if we did. Finally we gave up and closed the valves for good on the midship head. We peeled off into the current, swung downstream, and headed for salt water.

The Hudson was broad and misleading, the surface smooth, as if no current flowed at all. But buoys wobbled violently on the surface, whipping back and forth like balloons strung in a windstorm. At bends in the river the surface might make a smooth turn, then erupt with sudden bulges of turbulence as big as parking lots.

That evening, the river gleamed with sunset colors. To the west the Catskills were torn-paper silhouettes against the sky. Hills along the river grew higher and the homes on the bluffs statelier. A few turreted mansions rose above the trees, and orchards and hayfields covered the hills in undulating rectangles. Thomas Wolfe wrote somewhere that the Hudson River reminded him of "old October." That was it, exactly. I saw it too. An old river, the color of clay, its surface reflecting earth colors.

I thought of the English explorer Henry Hudson, who in 1609 became the first European to ascend the river. In his ship, *Half Moon,* he sailed about 150 miles upstream, to present-day Albany, hoping to find a route to the Pacific and claiming all the land he saw for Holland.

After dark, the river got tricky. High water made the current so strong that boats laboring upstream made scant progress, while we, hurtling downstream, seemed to fly. We made eight or nine knots, and could not go slower. We were forced to watch constantly for debris—the usual logs and whole trees, but also picnic tables and park benches and loose sections of dock. Somebody had to stand in the bow at all times, scanning the water with a spotlight and making sure Hajo was aware of any large objects ahead.

Debris was not the only challenge on the Hudson. Marker buoys are mysteriously rare on this broad river with its ocean-vessel traffic and numerous shoals. Even some of the buoys that were marked on the charts were gone, probably washed away by high water. The wind was too strong to keep the charts on deck, so we spread them on the table below in the main cabin, where we could read them and keep Hajo notified of upcoming landmarks.

Every time we passed a buoy we hit it with a spotlight, its reflective numerals bursting into view. If it was the buoy we expected, it meant we were on course, with enough water beneath our hull. But a mistake could carry us onto flats only a few feet deep.

By eleven o'clock, Paul, Harold, and Tim had to get some sleep—they would be rising at 4:00 A.M. for the morning watch. Hajo, Matt, and I stayed on deck.

From the bow I would see something unexpected dead ahead and hurry back to Hajo at the helm and ask, "Is that a bridge?" We would raise our binoculars to examine the lights and yell down to Matt in the cabin and ask him to check the charts. He would yell back, "It's a bridge all right, how did we miss that?" We would spend the next fifteen minutes as we closed the gap trying to spot the green buoy that was supposed to be on the upstream side of the bridge and the red buoy that was supposed to be on the downstream side. But they were nowhere to be seen, and we ended up passing beneath the center span on faith alone, checking the depth finder every five seconds and hoping no barge came chugging around the bend pushing its five-foot bow wave and steered by a helmsman half-asleep and not caring a damn about smaller craft anyway.

All night it was like that. The hours passed quickly, and land slipped silently past. I stood alone in the bow, in the darkness. Starlight shimmer made the river a fluid mirror, and the darkness was filled with fragrance and wild sounds. Even above the engine I could hear the racket of frogs and peepers in marshes a quarter mile away, crickets sawing away in the woods beyond. All the night creatures had fallen into a bleating cadence, synchronized, maybe, to the secret panting machinery of the earth. Amazing how smoothly nature's engine runs, though that night it sounded like some of the bearings and rocker arms needed greasing.

I woke Harold at 3:00 A.M. as we approached West Point, too exhausted to stay on deck any longer. Hajo remained at the helm, on his way to a twenty-seven-hour marathon.

I slept hard until seven-thirty and climbed on deck to find the river in lovely calm, dressed in haze, with the Palisades rising like Montana around us. Blue sky and clouds over the cliffs and the thick rounded forests above them—it was like slipping into a painting by Thomas Cole. Or almost. The river was lovely, but it was not the same river glorified by the Hudson River school. That river never existed. Cole and his contemporaries were intent upon making the Hudson the Rhine of the New World and they lit it in splendor, as it might have been imagined by angels. To them, the idea of the river was more important than the river itself.

In quick succession we passed Dobbs Ferry, Hastings-on-Hudson, Yonkers, the Bronx, Riverdale. We passed the mouth of the Harlem River. We motored beneath the George Washington Bridge and admired Jeffrey's Hook Lighthouse in its shadow—the "Little Red Lighthouse" of the children's book by Hildegarde Swift. The lighthouse was scheduled to be dismantled when the bridge was built in 1931, but children who loved Swift's book mobilized to save it.

Everyone was on deck by now. Paul, Harold, Tim, and I had never seen the city from the Hudson and couldn't get enough of it. We exclaimed over midtown and the Empire State Building and the World Trade Center—and of course we could not have imagined the fate awaiting those twin towers fifteen months in the future, and the consequences of their destruction. We were happy in our ignorance. We snapped photos like yokel tourists.

I had never realized what a *river city* New York is, with its canyonlike streets culminating at piers on the Hudson and East River, every building designed to support and complement the waterfront. Navy Pier and Pier 40 seemed to be the real New York. Midtown was a cluster of buildings on the Hudson; lower Manhattan an elaborate development best accessed from the upper bay. The little New York Waterway ferries shot across the river as if they owned it. We smelled salt in the air.

Hajo and Matt had seen it all many times before. To them, the city was just another navigation problem. It required them to remain alert for ferries blasting across the bow and pleasure boats planing past and massive cruise liners inching up the river from New York Bay. Hajo and Matt maintained a professional detachment until we swung around Battery Park and entered the East River and they spotted wooden masts above the piers. Instantly, they jumped to their feet and ran to the port side, binoculars in hand, and tried to identify the tall ships docked at the city.

"Is that *Ernestina?*"

"No, I think it's *Aurora.*"

"There's *Peking!*" Hajo shouted, pointing to a large vessel docked at the South Street Seaport Museum. He turned to us and explained that *Peking* was a 377-foot, four-masted barque built in 1911 to haul cargo from South America to Europe, via the treacherous waters around Cape Horn. The legendary filmmaker Irving Johnson booked passage on her in 1929 so he could experience a Cape Horn storm. On that voyage he shot the documentary *Rounding Cape Horn,* which includes the most dramatic footage ever filmed of a crew fighting for their lives on a square-rigger. Johnson later wrote about the experience in his book, *The Peking Battles Cape Horn.*

We passed beneath the Brooklyn Bridge, the Manhattan Bridge, and the

Williamsburg Bridge. Ahead was Hell Gate, which Hajo had been worried about since we left Michigan.

Hell Gate is the turbulent junction of the East River with the Harlem River, near 92nd Street on Manhattan's Upper East Side. Hajo claimed it was one of the most dangerous spots on the East Coast and was infamous among sailors around the world. Turbulence in the Gate is created by narrow confines and conflicting currents—both river and tidal—compounded by barge and tug traffic. It can be hellish. The name was bestowed in 1614 by Dutch explorer Adriaen Block, a navigator and cartographer hired by merchants in Amsterdam to explore the Hudson River and establish a fur trade. After wintering near Albany, Block descended the Hudson and followed the East River Passage to Long Island Sound. Block had already been forced to build a new ship after his first caught fire and burned. He nearly lost the new vessel in the turbulent junction of the Harlem and East rivers. He called the maelstrom there "Hellegat"—Hell Gate.

When the tide is flooding, the currents at Hell Gate create standing waves ten or more feet in height. The difficulties of the heavy water are magnified by the proximity of Mill Rock, Rhinelander Reef, and Heel Tap Rock.

But Hajo had planned carefully. With Paul at the helm we entered Hell Gate at slack tide, while it was in a relatively peaceful mood. Even then, the current was strong enough to turn the boat sideways and nearly throw us into the rocks and broken concrete riprap along shore. Hajo grabbed the helm with Paul and both men wrestled the boat under control.

We passed through the turbulence and entered a channel lined with destroyed marshes under the flight path of La Guardia, past Hunts Point and Rikers Island, past an abandoned and listing prison ship and old tugs and freighters rusting in the muck of side channels, and a small, overgrown island capped with the charred stone chimney of what was once someone's mansion at the edge of Long Island Sound.

Then we were in the Sound and I was surprised how much like a Great Lake it seemed. Smaller, and of course salty. But capable of fury worthy of Superior.

Chapter 15

THE ATLANTIC

Long Island Sound ◆ New Crew in New Haven ◆
A Storm on the Sound ◆ On to Maine

We arrived in New Haven Harbor after sunset, while the sky was still washed with scarlet and the lights of the city were just coming on. This was home port for Hajo—he lives a dozen miles down the shore at Indian Cove—and he hadn't been home in three months. "This is my harbor," he said, excitement in his voice. "See that lighthouse? That's my lighthouse. See that sunset? That's my sunset."

We lowered the yawlboat and I gave him a ride across the harbor to Oyster Point Marina, where we tied up to the public dock and walked across a parking lot lit by lampposts. Hajo's wife, Lee, sat waiting in her car. She gave him a wry smile, as if to say, "Well, *there* you are," and he grinned and shrugged a kind of apology. He got in on the passenger's side and they drove off.

Matt and I would pass the weekend on the *Malabar*, at anchor in the harbor. Harold and Paul were leaving in the morning, as planned. They would take the train from New Haven to New York and catch planes at La Guardia. Tim announced that he would be leaving then also. He was already packed.

In the morning the four of us piled into the yawlboat and motored to the marina, where we showered and did laundry. The beach between the docks was covered with oyster shells, remnants of the industry that made New Haven the largest oyster producer in North America. In the late 1990s, a waterborne disease reduced the stock of shellfish throughout Long Island Sound, decimating the New Haven fishery. Only a few oyster trawlers still work the harbor and offshore waters. We watched them motor out in the mornings and return in the evenings, gulls swarming above their open decks.

Matt and I gripped hands with Harold, Tim, and Paul. Already we'd exchanged e-mail addresses and phone numbers. Tim and Matt weren't speaking, but the rest of us made promises to get together sometime.

Hajo rejoined Matt and me for dinner that night and we feasted on oysters and lobsters, but it was a subdued meal. The boat seemed suddenly very large and empty. Hajo promised we would eat only seafood the rest of the way to Maine. He had to go home for the night, but he promised to return the next day with new crewmembers. He'd been calling some people he knew. It wouldn't be a problem.

At anchorage, Matt and I had few opportunities to relax. Hajo was worried that the tide and wind would cause the *Malabar* to swing, wrapping her anchor chain and dragging the anchor, so he had left us with specific instructions to take turns standing anchor watch all night. We tried two hours on, two hours off, but kept falling asleep on our watches. Finally, we just gave up. For the next few nights, whoever happened to wake up would climb to the deck and check the anchor. Once or twice the wind was strong enough to make the boat swing in broad sweeps at the end of the chain. But the anchor held and we quit worrying about it.

Hajo came to the boat Sunday afternoon in a bad mood, the worst I'd seen. He had just gotten off the phone with Steve, who wanted to know why we were still in New Haven. Steve claimed we were behind schedule, that our delays were costing him money, that bad weather was forecast for Wednesday, a nor'easter with gale-force winds, and we had to get out ahead of it. The message to Hajo was clear: Sail now or you're fired.

The only good news was that Hajo had found three crewmembers and brought them along in the yawlboat. Matt and I met them as they passed their bags up to us and climbed aboard. Stan Baldyja was an old friend of Hajo's from Suffield, Connecticut, who had sailed the East Coast all his life. He was a wiry fifty-one-year-old with a somber set to his face and seemed intent on living up to the stereotype of a crusty New Englander. It would be two days before he said more than two words to me. The words were "Yape" and "Nope."

Also coming aboard were a large soft man in his forties I'll call Bert and a young woman, sixteen or eighteen or twenty, who I assumed was his daughter. Soon I would see him put his hand on her thigh in a most unfatherly way. Let's call her Dolly. She had braces on her teeth and wore shorts that accented her baby fat. Parts of her spilled rather spectacularly from a halter top. Hajo had met them while he was getting gasoline in the marina. They were living aboard Bert's 37-foot sailboat and trying to save enough money to get to the Bahamas for the winter. Dolly had a job at a Dunkin' Donuts up the road, the 4:00 A.M. to noon shift. She was excited because she had just been promoted to assistant manager. Bert was elusive on the subject of what he did for a living. Nothing, I think. They climbed on deck, their eyes bright with adventure. "I can't believe we're doing this," Bert said. "I know," said Dolly. "This is just so outrageous," said Bert. "I know," said Dolly.

Steve's reprimands over the telephone had made Hajo angry. We would leave that evening, he said, regardless of the weather forecast. And the forecast was not encouraging. Thunderstorms. Winds fifteen or twenty knots from the northeast, the worst direction for a vessel bound for Maine.

"Let's wait until morning, see what the weather brings," Matt suggested.

"Get this boat ready to sail," said Hajo.

Late in the afternoon we watched a thunderstorm bear down on New Haven. It was preceded by a blackish-green storm wall that swept eastward, lancing the Yale campus with lightning. The center of the storm missed us, but a squally wind blew flurries of whitecaps across the harbor and made our boat strain against her anchor chain. After a few minutes the wind switched directions from southwest to northeast, forcing Hajo to start the engine and put the boat in gear, steering into the wind at half-throttle to keep her from dragging the anchor. Then came rain, hard, in driving sheets. When it passed, the harbor went calm and blue sky showed through cracks in the clouds. We raised the anchor and got under way. By six-thirty we were in the Sound, raising sail to catch a brisk wind from the south, heading for the open ocean.

But soon another storm began building behind us. It was much larger than the first. It climbed above the horizon until it was as wide as the sky, and came toward us fast, with nothing behind it but blackness. It was like watching a lid get pulled over the world.

We lowered the sails and double-lashed them to the booms. We pulled on our foul-weather gear, then battened down or stowed whatever was loose and stacked the washboards to seal off the companionway hatches.

By now, the entire northern half of the sky was filled with storm, the frontal wall moiling, the black interior lashed with silent lightning. We would later hear meteorologists describe it as a large, fast-moving depression that engulfed much of New England and surprised everyone with its size and power. It was circling counterclockwise, as storms always do in the northern hemisphere, sucking surrounding air into it and creating conflicting winds. We were in the lower, right-hand corner of the gyre, the "dangerous quadrant," in sailors' lore.

Hajo stood at the helm, yelling into his cell phone, trying to get an update relayed from his wife on shore. We thought the storm might pass us to the west, so our best tactic was to continue heading east. The National Oceanic and Atmospheric Administration (NOAA, or "Noah" as it is universally known) was still reporting only a chance of scattered thunderstorms and waves two to three feet. In another hour, long after the storm was upon us, a no-nonsense voice from the Coast Guard would blare over the radio, warning all boaters to seek sheltered anchorage immediately. By then it would be too late.

The wind grew stronger by the moment. It shifted, backing from south to east. Thirty seconds later, it swung to the north. The changes were abrupt and alarming, and each brought increased velocity. The new wind blew the tops off the waves, raising a horizonal spray that raced us downwind. Waves became confused, running south and north at the same time, slamming together and clapping spouts that the wind stripped away in banners. *Malabar* jumped like she'd been jabbed. We went from six knots to eight, with nothing catching wind but the bare sticks and the poop deck.

But we couldn't go fast enough to keep ahead of the storm. The black lid closed over us. Darkness fell and the air temperature dropped ten degrees. Raindrops the size of ball bearings hurtled down. Hail bounced across the deck, and a roaring wall of rain followed. It fell with such intensity that it flattened the tops of the waves.

All hell broke loose. Lightning flashed, a dozen strikes at every moment, in every direction around us, thunder cracking and rolling continuously, with instantaneous teeth-rattling detonations so close we could feel their impact. It was a barrage; we were being carpet-bombed.

I pulled the bill of my hat down to my eyes and cinched the rain hood as tight as it would go. Still the rain blinded me. My face ran with salt water—I couldn't tell if it was the sea or my perspiration—while more lightning than I have ever seen strobed the air around us. It filled the night with stop-action images: waves poised and sharp-crested, four- and five-footers now; the masts, tall and lightning-prone, leaning ten degrees to port, then to starboard, then to port; Stan standing resolute in his yellow slicker, his legs apart for balance, both hands on the helm, staring ahead as if he could see through the rain and the night-vision-destroying explosions of lightning. The wind hollered, and waves burst against the hull. I heard the hiss and rush of the boat cutting through the water, the dismaying random banging of our rudder loose in its fittings. Over and around every other sound was the continuous, rumbling, brawling and booming of thunder. The loudest peals were nearly simultaneous with the lightning, and seemed to fill me from the inside out. I kept shutting my eyes, expecting blows to the head.

At some point Hajo yelled at me that I should go below and try to sleep. It would be a long night, and I had volunteered for graveyard shift, 1:00–5:00 A.M. Hajo would not sleep at all, and it would be better if I was rested.

I went to my cabin and laid in my berth, fully clothed, gripping the sides of the bunk to keep from falling out. Lightning filled my cabin window. I could feel the seas building beneath us.

Of course it was impossible to sleep. I treated it like duty, clenching my eyes closed, willing myself to rest. But every rise of the boat on a wave was followed by a stomach-yanking drop and the crash of the hull against water.

The boat pitched, rolled, and yawed. I waited for signs of seasickness and was surprised to find none. A month ago I would have been vomiting by now.

I felt trapped in the tiny room. Something kept banging deep in the boat. I wondered if it was the rudder, about to break, or the drive shaft, or the masts, flexing, splitting the boat in half.

Water roared against the hull only a few inches from my head. I could feel the boat shouldering into the waves. I thought of bridges swaying until they reached a critical oscillation and collapsed. I imagined the rudder breaking off; I saw the stem, weakened with dry rot, falling away and carrying the bowsprit and headgear beneath the hull and fouling the propeller; I saw the engine compartment flooding, and the engine stalling. Worst of all, I saw the hull disintegrating. Hajo and Matt sometimes called the boat "The Rock," an affectionate reference to her concrete hull. But concrete sinks—Christ, it sinks like rock. In March, when they had arrived in Michigan to begin work on the *Malabar*, they had found her hull in such bad shape they had punched holes in it with their fists. They had spent twelve weeks repairing the damage, but only above the waterline. They had no way of knowing how bad the damage was below, which was why we checked the bilges so religiously. But what if she was as rotten beneath as she had been above? What if the entire structure fractured under the stress of these waves and the concrete crumbled the way concrete buildings in the Ukraine crumbled to rubble during earthquakes?

I imagined the hull bursting open, the boat nose-diving into a wave and not coming up. She submarined into green darkness, clouds of bubbles surging behind. She did not stop until she hit sand three hundred feet down.

I had to avoid such thoughts. I got out of my bunk and was thrown against a wall. Better to go up top and try to help. I was thrown against another wall. Better to stay busy than remain in the cabin and torture myself.

On deck, the scene was grim. In the lightning flashes I saw Hajo standing at the helm in his foul-weather gear, both hands fighting to hold the wheel, legs braced wide on the deck. Bert stood on one side of him, Stan on the other, both leaning against the rails and staring ahead through their drawn hoods. The rain came in a steady downpour, and the waves were bigger—rollers and chop mixed, a complexity of forces. The biggest rollers, eight-

and ten-footers, came now from the stern, and *Malabar* yawed as she descended into the troughs, making six or seven knots on the downside, then slowing to four or five as she climbed the next wave and met the wind, thirty knots now. Lightning flashed around us, but it seemed to be moving away.

I clipped onto a safety line and staggered to the stern. "How's it going?" I shouted to Hajo. I had to lean close to hear his answer.

"We turned around!"

"What!"

"We're going back to the harbor! We almost got to the Race, but the seas were too big and we couldn't control the boat!"

The Race is the milewide channel at the mouth of Long Island Sound, where the Sound bottlenecks between reefs around Fisher Island and the mainland. Here tidal currents converge from Long Island Sound, Fisher Sound, and the open Atlantic. The worst time to be there is when wind-driven waves meet flowing tide. Their opposition creates turbulent, choppy seas of the sort everyone fears.

It had not been an easy decision to turn back. Steve would be angry. But Hajo had to put the safety of the crew first. Now he was furious with himself for capitulating to Steve and setting out from New Haven in the first place.

The compass, its glass running with rain, was lit by Matt's red penlight clipped to the binnacle. The batteries ran down every hour and we would navigate blindly for a few minutes while Matt fumbled in the darkness to change them. Even with the compass lit, Hajo had trouble reading it.

"How's your night vision?"

"Good."

"Take the helm."

So I held *Malabar* by the reins and she ran, engine chugging mightily, rising on each swell, yawing hard to port as she rode it into the trough—the entire length of her twisting and flexing before me. It was like steering a hundred-foot wooden sled down a slope. Once she started turning, she kept turning. I steered more by will and hope than by mechanical process, zigzagging down the waves.

The compass swung wildly, ten to twenty points every wave. "Watch your course," Hajo suggested when he realized how far we were deviating.

I tried, but it seemed hopeless. The best I could do was keep a rough average between northwest and southwest.

The wind changed again, this time from northeast to south, throwing the already confused seas into deeper confusion. A powerful set was pushing us toward the Connecticut shore. Even if we stayed on a perfect course, straight toward the west, we might still be driven onto rocks.

Past Faulkner Island Lighthouse the wind shifted again. It came from the north now, and stronger than ever. The seas struck the starboard side, and the scuppers flowed. Bert, standing at the rail trying in vain to see buoys and lighthouses, wore a rain jacket but had neglected to bring waterproof pants and sea boots. He wore corduroy slacks and topside loafers and was drenched to the skin. Spray washed over him, a drencher every thirty seconds, but he never complained.

The wind veered suddenly to the east, the fifth significant change, and now the wind and waves both came from the stern and *Malabar* put her butt down and scooted, the water hissing in her wake. All night the rollers from the open Atlantic had been building. Now they were joining the chop and making bigger waves than we had yet seen. I thought of concrete chunks breaking off and sinking into the ocean.

We took a series of ten-footers over the sides. I'm being conservative. I stand just under six feet tall and at the helm was elevated an additional six feet above the waterline, so I had some perspective. The rollers came at eye level. One wave threw the boat over so far that dishes in the galley broke free and crashed to the floor. A fire extinguisher skittered past Matt where he leaned over charts on the table. He tackled it and lashed it to a stanchion.

Through the rain we could see shore lights, then, unmistakably, the lights of New Haven. We lined up on what we thought were the green and red buoys marking the channel entrance, but they turned out to be markers on the breakwall behind. Visibility was inconsistent—a mile one minute, a hundred feet the next. Prominent in New Haven were a pair of range markers, dazzling blue beacons that line up in accord when you get your boat straight in the channel. Hajo took the helm, and I stepped aside with relief.

The confusion over the red and green markers had carried us too far; we would need to swing the boat east a quarter mile to line up properly with the harbor entrance. It was a critical moment. Hajo waited for a wave to pass, then spun the helm with both hands as fast and hard as he could and the boat turned broadside to the wind.

A wave came at us, rising out of the darkness. It was bigger than any we'd seen that night. A twelve-footer, maybe fifteen. It hit us broadside and the boat started going over—twenty degrees, thirty, perhaps forty degrees, until the masts seemed about to splash into the sea. Everything loose in the cabins below went flying. We could hear glass breaking and heavy objects tumbling. We hung on to lines, rails, samson, and helm, and for a moment of black horror it felt like we were going all the way over.

But she righted herself. She swung almost but not quite as far to starboard, then to port again. But she had her legs beneath her now and we were running toward the breakwater and the narrow entrance to the harbor. The range markers lined up. Hajo would say later it was the boat, not he, who steered us to safety.

By four in the morning we had dropped the hook in our old spot in the harbor and had the generator gassed and running. All of us sat around the table in the main cabin, drinking whiskey and beer. We were wet and shivering. Dolly, who had spent the entire voyage in her berth, seemed to be suffering from mild shock. She appeared younger than ever, a wide-eyed little girl who had seen too much of the world too soon. Bert glanced at his watch, then at Dolly. "You've got time to get to work," he said. She looked at him in disbelief, her eyes wider than ever. "Why not?" he said.

We lowered the yawlboat and Bert and Dolly climbed down the ladder into it. Matt tossed them their bags. "We'll be back after noon," Bert said. "Don't leave without us. We really want to go to Maine with you guys."

But of course they went to shore and we never saw them again.

The rest of us stayed up another hour, not talking, just sitting in the cabin drinking. No camaraderie filled the room. We were alone with our thoughts. Stan, who still had said nothing to me except "Yape" and "Nope," kept holding his glass out to Hajo for refills.

I don't want to seem overly dramatic. Worse storms occur all the time. It could not compare to the blow that sank the *Edmund Fitzgerald*, and of course it was nothing compared to what the Atlantic is capable of.

But we had glimpsed what every survivor of every storm has seen, and we understood a little of what they've come away knowing. We'd confronted the blank face of the sea, and it was terrifying. Behind it was indifference so profound it crushed all hope. We were fragile and temporary, and could be annihilated at any moment with no more effort than it takes to crush a bug beneath a heel. The world itself was lethal, and God was elsewhere. We were utterly inconsequential.

Hajo poured whiskey until his glass brimmed. "We could have died out there," he said.

"Damned straight," said Stan.

In a few days the weather cleared, and we set off again. I remember the *Malabar* motoring over the Race in bright afternoon sunlight, passing from the greenish-gray water of the Sound to the greenish-blue water of the open ocean. We caught an outgoing tide and rode on a river a mile wide and two hundred feet deep, filled with powerful upswells and dangerous swirls and choppy whitecaps. It was mesmerizing; I'd never seen a place like it. I was glad to be seeing it in daylight rather than at midnight in a storm.

We were a crew of five again: Hajo, Matt, Stan, and me, plus Hajo's sixteen-year-old son, Luke. Luke was a student at the Sound School in New Haven, a private high school that emphasizes maritime studies. The students there spend a good portion of the school year on the water, often crewing the schooner *Quinnipiack,* and devote much of the rest of their time to biological studies in the field and to building boats. Luke seemed to take for granted that the only life for him was the life of the sea. He was relaxed and confident. He called his father by his first name and seemed amused by the old man. "Hajo is sort of a dufus, sometimes," he said, sounding like any sixteen-year-old. "But he's okay. Yeah."

In the open ocean off the coast of Rhode Island we encountered fog so dense we could barely see from one end of the boat to the other. Our eyes ached from staring through it. A songbird—a yellow warbler of some kind—tried to land on the mainboom, fluttered, almost landed, fluttered again, suffered a failure of nerve, and flew off into the whiteness.

We couldn't see far, but we sensed the vastness around us. Long-period swells heaved beneath the boat; they had the feel of continents in them. We made our way by compass and GPS, from way point to way point. Every buoy emerged from the fog exactly where it was supposed to be—a triumph of satellite technology. Boats signaled with blasts of their foghorns. We used a handheld one, of the sort popular with fans at football games. We blasted, others blasted back, then suddenly a trawler or recreational troller or lobster boat emerged through the dissipating whiteness and passed to our port or starboard. People on board watched us with mild interest, sometimes waving but not always.

Out of the fog, precisely where the GPS said it would be, came a buoy moaning with fog warnings, then a lighthouse perched on a shore of sand, beach houses, a rocky point, and we entered the narrow channel to Big Salt Pond, the bay at the center of Block Island. On the shore an old man, bent with years, walked a golden retriever on a leash.

Later, I fished with Luke in the channel, the *Malabar* anchored somewhere out of sight in the fog. For bait we used oysters Luke had bummed from a restaurant in town, gobbing the slimy meat on treble hooks and casting them with big bell sinkers that splashed in the water and sank instantly to the bottom, where fluke live. But the tide was slack and fluke feed best when the tide is ripping, carrying food over bottom where they lie flat and camouflaged, watching with their Picasso eyes for the chance to grab an easy meal.

We watched spider crabs and green crabs walking self-importantly along bottom in the shallows. Two horseshoe crabs were mating, the smallish male riding the female boldly from behind. The water was clear, the sand clean, the beach stones bright with colors. It could have been the Lake Michigan shore, except for the lobster shells on the sand and the scent of salt in the fog. On the dunes were thickets of wild roses identical to those along the Great Lakes, but twice as tall.

In the morning, with the wind out of the southwest—the sou'wester we'd been looking for to push us downeast to Maine—we ignored the NOAA forecast and left the shelter of Block Island. The forecast was for fog early, then waves seven to ten feet. We dreaded the thought of sailing into Buzzards Bay in 25-knot winds, but we were anxious to go and had become skeptical of weather forecasts. Sure enough, the wind stayed moderate, the waves two to three feet, and the day turned out blue-sky bright and so warm we stripped to our T-shirts. We raised the mainsail and jury-rigged a spinnaker, and the *Malabar* pitched and rolled in a kind of prance. Hajo said that she seemed happy to be in salt water again.

We sailed into Buzzards Bay, into blue-green water brilliant with whitecaps, the ocean rolling beneath us. Astern, the sun lit the waves the color of hammered steel. Stan stood at the helm and told me that the oldest written mention of sailing was a fragment of ancient Arabic text that read, "Sailing is victory." In the wind, with the *Malabar* surging through the sea beneath us, I understood. Victory over the wind, which can kill you. Victory over the sea, which can kill you a lot quicker. Over the boat, over the sails, over your own limitations. Sailing joins elemental forces in opposition, and lets you ride them. The illusion is created that you've harnessed the elements. You've thrown reins on a lion and leaped on its back. Of course the lion is ultimately untameable, and if you're careless it will turn around in a flash and eat you. Which is also the appeal.

Luke told a story about sailing in a tall ship race on the *Brilliant*, a gaff-rigged schooner based in Mystic Seaport. "We were off Newport," he said, "and the winds were awesome and we had all the sails up. It was like today, only a lot better—taking advantage of the wind, using every inch of sails. Yeah. And we weren't afraid to drag the rails in the water, man, we were making ten, eleven knots, it was sick, everybody just loving it, I mean those of us who sailed, mainly this other kid and me. I liked him, he was a black kid who cared about sailing and the ocean and shit, not like the other kids, who just wanted a diploma. All the others, like ten or twelve of them, took seasick medicine and like an hour later they were boom, asleep, all stacked up on the deck with their heads a foot from the rail so when they woke up they could hurl over the edge without having to move much. Even our teacher was hurling, man, it was cool, and this other kid and me were sailing the boat all by ourselves, just digging it, eight-foot waves and thirty-knot winds and the *Brilliant* just flying. Yeah. It was so sick."

Then we were shooting through the Cape Cod Canal at evening, with the sails up and a tail wind and a flowing tide. We made 10.7 knots, the fastest speed of the trip. We flew past people Rollerblading on trails along the canal and others fishing with surf rods for stripers, and catching them.

We entered the ocean beyond the canal, a fiery sunset to the west, while in the east an orange moon rose. As it climbed, night fell, and the moon's trail became a long canyon of silver across the sea, swallowing the darkness encroaching on each side. We pushed on all night, cutting across the Gulf of Maine, in open ocean.

The next day, miles beyond sight of land, the water was so blue and clear that I dipped a bucketful and tasted it because it seemed as if it must be as sweet as Lake Huron. I sat for hours in the bow and watched seabirds skimming above the waves, rising and falling with uncanny precision over the swells, always staying one and a half inches above the water no matter if they were in a trough or above a crest, darting away at the last possible moment to avoid the breaking tops of waves. I inhaled the ocean wind and walked the deck while long slow rollers ran beneath us—waves so different from the waves on the lakes that they seemed made of a different substance altogether.

I became aware of an alertness growing in me. The vast unfamiliarity surrounding the boat was a little like walking in a strange neighborhood at night. I wasn't frightened, just heedful. I remembered feeling a similar alertness the other times I'd been on the Atlantic—on a small boat off the coast of Iceland, in company with scientists and journalists, most of us seasick; and going to Newfoundland during a storm one night (and most of us were sick then, too) on an enormous car ferry that a few months later sank in a similar storm. Faced with an overwhelming unknown, we fall back on memories. I thought of Joyce's "scrotumtightening sea," and of Thoreau, on Cape Cod, claiming, "The ocean is but a larger lake," and tried to remember what Saul Bellow's Herzog said about it ("I fell in love with the Atlantic. O the great reticulated, mountain-bottomed sea!"), and what one of the Lowells wrote about its monotony (James Russell Lowell: "There is nothing so desperately monotonous as the sea, and I no longer wonder at the cruelty of pirates"). Of the authors Hajo adored, I had read only Melville, Conrad, and Slocum. I knew none of the grand ocean adventures Hajo had read and reread, or the sea captains' memoirs in their dense Victorian prose, or the jubilant ballads of John Masefield, which on occasion would burst forth from Hajo: "I must down to the seas again, to the lonely sea and the sky/And all I ask is a tall ship and a star to steer her by."

Maybe someday I too will break into song, but for now I could not get past noticing that everything, everywhere around me was the same. The waves were all the same. The birds that at first had seemed such a wonder never changed. Hajo had said that the biggest challenge to ocean crossings was the endurance it required. I believed him. You would have to endure the sameness of the waves—so monotonous that once or twice as I watched the horizon my heart leaped when I saw a single wave higher than the others, silhouetted against the sky: I thought it was land. Hajo, Matt, Luke, and Stan, sailors all their lives, were energized. The naked horizons, the long rollers and the whitecaps, the immensity of it all made them feel more alive. It was a greater freedom than they could find along the coast, where our passages are spoiled by other people and the many reminders of work and responsibility. I took the helm for a few hours and it helped. There was satisfaction in doing something useful, and I liked the sensation of riding the boat over the rollers. But an oppression weighed upon me. If I had thought it was difficult to know the Great Lakes, what hope did I have on the ocean? My inland sensibilities were overwhelmed. I would need more time, next time, to acclimatize.

I thought of Rudyard Kipling. He too was made uneasy by unfamiliar waters, but in his case it was the sea he knew, the Great Lakes that were strange. His impression of the lakes appears in *Letters of Travel, 1892–1913*: "There is," he wrote, "a quiet horror about the Great Lakes which grows as one revisits them. Fresh water has no right or call to dip over the horizon, pulling down and pushing up hulls of big steamers, no right to tread the slow, deep sea dance-step between wrinkled cliffs; nor to roar in on weed and sand beaches between vast headlands that run out for leagues into bays and sea fog. Lake Superior is all the same stuff towns pay taxes for [fresh water], but it engulfs and wrecks and drives ashore like a fully accredited ocean—a hideous thing to find in the heart of a continent."

At midnight of the night of the full moon we came at last to the coast of Maine, a barren rocky silhouette against the sky. It had been exactly one lunar month from the start of the trip on Lake Michigan. Now the moon stood straight up in the sky and lit the waves to shimmering silver. A warm shore breeze swung into our faces, carrying the fragrance of woodsmoke and evergreens, and Hajo turned to me and said, "There! Smell it?" It was the smell of land—rich and sweet and complex after the pallate-cleansing winds of the open sea.

We would anchor for a few days among the lobster boats in Bar Harbor and go to shore frequently to explore the town and the island. One night Luke and I would take the yawlboat out in the harbor and he would show me how to stir the water with an oar, activating bioluminescent dinoflagellates that bloomed in eerie green profusion with every swirl of the oar. A universe of green planets and stars danced in slow orbits around us.

Stan would tell me that he had an advanced degree in physics and had worked for years in the aerospace industry as a metronomist, a scientist of time. His specialty was figuring out ways to use time efficiently, measuring intervals of as little as ten-thousandths of a second during the construction of space suits. When he was younger he had joined the Peace Corps and was sent to Indonesia, where he had the incongruous assignment of teaching physics to tribesmen on the island of Bali. They were attentive students, he said, very curious about the mysteries of the universe. After class they sometimes led him into the forest to hunt for wild boars with spears. Now he worked in slow time, raising trout in ponds on his farm, reading the classics for recreation, and sailing whenever he could.

We would meet Steve Pagels, who seemed insufficiently appreciative of the *Malabar*. After a curt inspection and without a word of thanks to us for bringing his boat home, he would set his employees to work painting the cabin roofs, covering the unfinished teak with industrial yellow.

We would drydock the *Malabar* at the Hinkley Boat Yard in Southwest Harbor, lifting her free of the water in an immense sling, and have the satisfaction of seeing that her hull was sound beneath the waterline. She would easily earn her Certificate of Inspection from the U.S. Coast Guard. In two or three weeks Hajo and a new crew would sail her to Greenport, Long Island, and put her to work doing day charters on Long Island Sound.

For those few days we would be the bad boys of the Bar Harbor fleet, hosting parties, running the yawlboat back and forth from shore with loads of food and beer. Matt would fire his cannon across the bow of Steve Pagels's four-masted schooner, *Margaret Todd,* as she sailed past with a cargo of well-behaved tourists lined up on benches along the rails. The retort—a blank charge—echoed between the hills around the harbor. Steve was displeased, but his passengers leaped to their feet and cheered. They wanted to be on our boat. So did Steve's crew. They were college students, most of them. At night, after they were off duty, four or five at a time would row out to sit on deck with us and listen to Hajo's stories.

We would throw a farewell party on the drydocked *Malabar,* while she hung suspended in her sling and could be reached from the ground only by an extension ladder. Hajo baked an enormous stuffed turkey in the diesel oven and we grilled a dozen chicken breasts marinated in Luke's secret sauce (the critical ingredient was Sea Dog beer). We invited Steve's crew to join us. It did our hearts good to feed the poor kids. Steve worked them so hard and paid them so poorly that they seemed on the verge of starvation. Half a dozen of them were girls, ages eighteen to twenty-two, who filed one after another down the ladder to the main cabin and dove for plates of turkey and chicken, devouring fist-size hunks of meat that left their chins shiny with grease. All vegetarian inclinations were forgotten. While they ate, they watched Luke. They couldn't take their eyes off him. They struggled to understand the attraction they felt for this remarkably handsome, remarkably self-assured young man, who was clearly jailbait. It confused them. They watched open-lipped, with flushed faces, and glanced away guiltily when he caught them looking. Finally, we turned up reggae music so loud that objects began bouncing off shelves, and the girls grabbed Luke by the hands and danced him around the cabin.

We geezers watched, saddened at first, then laughing in the face of the injustices and humiliations of life. Oh youth! Stan caught my eye and grinned. Hajo jumped to his feet and cried: "We're *alive!*"

Chapter 16

LAKE MICHIGAN

Home Again ◆ Taking Issue ◆ Reflections on the Future
of the Great Lakes ◆ Swimming the Wreck

And so it was time to go home.

Stan and I rented a car in Bar Harbor and set off for the highway, took
a few wrong turns, found our way again, and joined a pod of grim law-
breakers hurtling at seventy-five miles per hour down Highway 95, bound
for the great metropolises. Along the way, the formerly reticent Stan spilled
his guts. It was a gift to me, and I'm grateful. We took the turnoff to
Manchester, New Hampshire, shook hands solemnly at the airport, and
vowed to stay in touch.

Inside the quaintly crowded terminal, I was elbowed out of the way by
a suited tycoon of twenty-four who was determined to get in line ahead of
me. Around us, every third person talked into a cellular phone. All were
doing more talking than listening, and every conversation was the same—
where I am, where I'm going, when I'll arrive, did you send the contract,
the deal's done, the deal's dead, we should get together for a beer. The
world had gotten on quite well without me. It does that.

As my plane descended into Chicago, I looked down at Lake Michigan
and thought how strange it was that such a large body of water could be
fresh. I remembered a newspaper story a few years ago about a young man
who was visiting Chicago for the first time and went out on the lake in a
personal watercraft. He ran out of gas and drifted away from shore and was
not rescued for a couple days. When he was found, he was so dehydrated

that he had to be hospitalized. A doctor asked him why he didn't drink water from the lake. The question startled him. He'd assumed it was salt water.

The Great Lakes hold a fifth of the liquid freshwater on the surface of the earth. It's an extraordinary fact. And an extraordinary responsibility. I'd never known it so surely.

In the months to come, I would hear several times from Hajo. He chose a crew from among Steve's employees and with Luke as first mate sailed the *Malabar* to Long Island. They had engine trouble along the way and encountered a storm and saw whales leaping clear of the water and landing in magnificent explosions of spray. "I wish you could have been there," he said. "It was magical." After a few weeks, he left the *Malabar* and went to work in the Gulf of Mexico as the captain of a research vessel.

Matt and Lisa got married in Baltimore, and Matt was hired as first mate on a tall ship sailing the Chesapeake.

Harold and his wife spent the summer in a campground on the north shore of Lake Huron.

Paul Garrett married his fiancée, Melissa, and they moved to southern Michigan. He continued to work on the *Saturn*, all the while dreaming of sailing the world as a rope chucker.

Tim published his third children's book, *Buck Wilder's Little Skipper Boating Guide*, and filled it with delightful illustrations of dozens of types of boats. The *Malabar* is there, manned by a grinning crew of four. No first mate can be seen.

Water levels in the Great Lakes would continue to drop and the shipping companies would continue to lose revenue, though by the middle of 2002, the lakes would be up a foot or so from their low in 2001. The perennial environmental concerns would remain in the news—alien invasive species, contaminated sediments and their effects on the food web, airborne toxins, urbanization, the usual conflicts between those who would protect natural resources and those who would use them until they're gone. In Michigan, Native American commercial fishermen would negotiate a revised agreement with the state that granted them perennial rights to fish with nets in certain portions of the Great Lakes. Sports anglers would protest, asking why Indians are allowed to net and sell fish that are raised and paid for with revenue from sportfishing licenses. The Indians would remind them that most of the native fish are gone, annihilated by alien species brought to the lakes by Europeans.

U.S. and Canadian biologists on Lake Erie would be astonished to discover large numbers of a small fish known as the silver chub suddenly reappearing in the lake. The species had long been scarce, and was assumed to be nearly extinct. Silver chub, *Hybopsis storeriana,* evolved in the lake to feed on small clams and native mussels. But they're extremely sensitive to pollution and very few of them had been seen since the degradations of the lake beginning in the early 1950s. Enough survived, however, to support a remnant population, and they apparently developed a taste for zebra mussels. With so much food available and the lake growing cleaner, the chubs were resurrected.

Not all the news from Lake Erie would be so optimistic. In the spring and summer of 2002, researchers would discover that the water throughout much of Erie's central basin was running out of oxygen. Fearing a return of "dead zones" similar to those that plagued the lake in the 1960s, teams of scientists from the United States and Canada converged to figure out what was wrong. This time the diminishing oxygen content was not caused by eutrophication, as in the past. In the 1960s, high levels of phosphorus— the result of runoff from agriculture and waste from municipalities and industries—had stimulated massive growths of algae. When the algae died it settled to the bottom, where it decomposed, devouring oxygen and making the water inhospitable to virtually all organisms.

But the new problem was more complicated. Phosphorous levels remained high, despite decades-long efforts to reduce their amounts, but without a corresponding increase in algae growth. With no obvious cause for the oxygen depletion, researchers were forced to pursue various hypotheses. Was atmospheric ozone depletion allowing ultraviolet light to penetrate deeper into the lake, warming the bottom and pushing away oxygen-rich colder water? Were petroleum, industrial chemicals, or unknown pollutants altering the water's chemistry? Were zebra mussels to blame? Were they consuming so much algae that a critical source of oxygen was eliminated? Were their feces and pseudofeces depleting oxygen as they decayed? Whatever the cause, the problem seemed to be internal to the lake, not simply a matter of shutting off sources of external pollution. One scientist would call the new crisis "the beginning of the second environmental war in Lake Erie."

In the politics of the lakes, two issues would gain increasing prominence: diversion of Great Lakes water and drilling for oil and gas. They became hot issues, especially in states where gubernatorial races were gaining momentum. Candidates who had never agreed on much of anything lined up on the same side. Public sentiment ran so strongly in favor of protecting the lakes that no politician dared oppose it.

Anyone aware of global water issues realized that the Great Lakes were becoming a commodity that must be protected from exploitation. The parched Southwest had already turned its eyes northward, and was gaining population and political power at about the same rate that the upper Midwest was losing it. As long as water in the lakes remained at low levels, nobody would dare try to sell it. The public would not stand for it. But the public's memory is short. The lakes will rise again, and when they rise high enough to erode shorelines and drag houses into the surf, as happened in the 1970s, then the public might well concede that a little siphoning of the excess is in order. That's when lobbyists for Texas ranchers will come calling.

Those of us who dream of five pristine lakes protected from commercial abuse will be wise to remember that local economies dependent upon the lakes are the final line of defense. The shipping industry, for instance, will always be among the fiercest defenders against water diversions. An industry that loses two hundred million dollars a year for every inch the lakes drop will not sit by and allow a foot to be taken off the top.

In a similar way, oil and gas drilling beneath the lakes is an emotional subject driven by economics. The public speaks up loud and strong against it, but its voice can grow quickly silent. When gasoline prices surged in the summer of 2001, many people who had been opposed to drilling started wondering if it made sense.

The poet Robert Hass and I once discussed the subject as we drove along the north shore of Lake Michigan. Hass is a former poet laureate of the United States and a professor at the University of California at Berkeley. He's a brilliant poet and a very smart man, and I value his opinion. I asked him what he thought about drilling for oil and gas beneath the Great Lakes. He said he thought it made sense. It made sense to reduce our dependence on foreign sources. It made sense to tap resources that are just sitting in the ground. It made sense to utilize available technology and put trained crews to work.

"But something can make sense," he said, "and still be wrong. If history has taught us anything, it's that there is never a shortage of practical, hard-headed people making one wrong decision after another because it makes sense."

Left to do what they want, people have always demonstrated an amazing capacity for making poor decisions. All the oil and natural gas beneath the lakes could probably run our automobiles and heat our houses for a few weeks or months. Its withdrawal could ease our dependence on foreign oil for a time and would earn a great deal of money for a few people. And in all probability, no harm would be done. If no pipelines rupture, if no well-heads burst or fires erupt or careless workers turn valves the wrong way or oil-filled ships run aground on rocks or trains run off their tracks, then history will judge the effort to be of slight importance and forget it. But if something goes wrong, future generations will condemn us. We'll be no better than those who decimated the herds of bison and leveled the old-growth forests and dumped poisons into rivers and lakes. We'll be another generation of exploiters who cared only for their own gain.

The time is long overdue to start using better judgment. With issues like petroleum extraction and water diversion, all we have to do is balance possible gains against possible losses. Tally the ledger sheets—then make decisions that will make sense a hundred years from now.

A hundred years should be nothing in the life of the lakes. In many ways, they've changed little since the last glaciers retreated ten thousand years ago. They are still broad and blue, still sweet with freshwater, still surrounded by sand beaches and rocky cliffs, still swept by storms so powerful that they erase most signs of human endeavor.

But of course the lakes have changed much in the last hundred years, and the greatest changes might be yet to come. What of their future? Will endemic species adapt to invasions of aliens? Will new ones invade? What will be the impact of continued industrial growth and increasing shoreline development? What about changes in climate? What if the waters warm ten degrees? Will the Great Lakes someday become the Aral Sea of North America—shrunken to a tenth their size, hopelessly polluted, surrounded by deserts of poisoned sand? Will people a century from now see a motherlode of clean water or a wasteland?

One of the consequences of the degradation of the environment is a vague but undeniable cultural despair. Not only are we ourselves temporary, but so is the planet we live on. But if there was one lesson learned in the environmental awakenings of the 1960s and 1970s, it was that we have a voice in how we alter the planet. We shaped a better future for the Great Lakes when we became outraged at presumptions of ownership. The lakes, like the oceans, are owned by no one and by everyone. They are among the few places left where we can go to be utterly adrift and free. They are town squares, and nobody has the right to foul them or sell them. When a few try to profit from what belongs to all, it is morally and legally wrong, and we have every reason to be angry. This is the water we drink—how dare they dump poisons in it? This is where we go to refresh ourselves—by what right would they pipe it to desert golf courses?

One summer afternoon we drove to Lake Michigan—my wife, my sons Aaron and Nick, and Nick's friends, Jeff, Dan, and Derrek. We followed a path through the woods, carrying coolers with sandwiches and mesh bags bulging with fins, snorkels, and masks. From the cool shadows of the pines we emerged suddenly at the edge of the lake. Colors burst before us—the lake and sky blue, the sand glaring yellow with sunlight. Good Harbor Bay stretched in a long arc north, all the way to Whaleback and Leland. Across the water sat the Manitou Islands. It was the view I grew to know so well during those six weeks in February and March, while studying the dynamics of sand and wind. It was the same water I sailed through on the *Gauntlet*, bound for the finish line at Mackinac Island.

In the foredunes, we paused to look at the way individual stalks of marram grass looped under their own weight until their tips touch the ground. Each had dragged a perfect circle in the sand.

We followed the shore, walking on the wet sand and splashing through the wash of waves. The wind came steadily off the water, blowing waves that curled and broke and washed up before us. Where the beach narrowed between the waves and the high sloping flank of Pyramid Point, we dropped our coolers and spread our blankets.

A few hundred yards offshore, a gleaming black object broke the surface. It appeared to be a rock, but I knew it was the boiler of a ship, the *Rising Sun*, a 133-foot steamer that sank in October 1917. The ship was owned by the House of David, a religious sect with a settlement on High Island, near Beaver Island. Bound from High Island to Benton Harbor with a load of potatoes, rutabagas, and lumber, the ship encountered a whiteout of snow squalls here in the Manitou Passage. The compass malfunctioned, and the captain tried to get through the channel on guesswork and intuition. But his luck was not good. The ship struck a rock off Pyramid Point, swung broadside to the waves, and began filling with water. Thirty-one of the thirty-two passengers and crew climbed into lifeboats and rowed through the breakers to shore. One boatload of women and children capsized, but all were pulled alive from the water. A few of the survivors, wet and nearly frozen, hiked up the shore until they spotted a lantern in the window of a farmhouse on Port Oneida Bluff. Pounding on the door, they alerted the farmer, who hitched a team of horses to his wagon and made several trips to the beach and back, hauling survivors to his home. Only one passenger was unaccounted for, an old man who had been asleep below deck when the ship ran aground. He awoke after everyone else was gone and spent the night alone on the disintegrating ship. In the morning he was carried to shore by Coast Guardsmen who arrived in a rescue rig hauled down the beach from Glen Haven by two teams of horses. The *Rising Sun* was abandoned. Eventually, she broke to pieces and sank.

The boys and I put on our fins and masks and waded out into the lake and swam. The waves swelled beneath us. We swam beyond the second sand bar, to where the water deepened and the swells surged around the old boiler.

It was black, made of curved plates of cast iron joined together by rivets the size of silver dollars. Waves rose against it and washed in eddies behind. We trod water for a few moments, resting. A wave surged and shoved Derrek against the iron. He lifted his hand in surprise. It was bleeding. Attached to the iron boiler were hundreds of zebra mussels, their shells as sharp as barbed wire.

We spit into our masks, rinsed, fitted our snorkels in our mouths, and dived. Ten feet down was bottom, cobbled with round stones and boulders, their colors bright in the clear water, sunlight dancing among them. As we descended, the water wrapped around us and squeezed. Half-buried among the stones were ribs of timber, a drive shaft, a scattering of sprockets, rods, and plates—the skeleton and entrails of the wreck, spilled across the bottom of the lake.

I reached out and touched a rib. It was stained black with age and hard as stone. I kicked with my fins and reached for the drive shaft. It lay across the bottom like a giant fossilized spine, black and gleaming, smooth to the touch, scrubbed by eighty years of sand and current.

We surfaced, cleared our snorkels, and dived again. Through our masks we studied the magnified details of iron, wood, and stone. We stayed down as long as we could, until our lungs ached, then kicked to the surface for air. Nick opened his palm and showed me a handful of bright gravel. He let the pebbles spill from his hand, took a quick breath, and jackknifed under again.

North American Indians have a myth, similar to one found in many cultures around the world, that the universe began as a vast sea inhabited by creatures that lived their entire lives in the water. But they grew weary of having to swim all the time. They wanted a place to rest. So one day Muskrat took a deep breath and dived to the bottom of the water, deeper than anyone had ever gone, and returned clenching a fistful of dirt. When he opened his paw, the dirt spilled out and became the earth.

The world overflows with bounty. It's rich and diverse and nearly inexhaustible. But here's the dilemma: The things of the earth are fascinating, amazing, bizarre, wonderful—and insufficient. We need more. We've always been driven to explore not only the physical world but our responses to it. Bracketed by mysteries, adrift, alone, despairing of our ignorance, we turn to the physical because there, at least, we can know a thing or two for certain. But we're creatures of spirit, too. It's our spirit that makes us encounter the wonders of the world and know that they are wonders.

The boys burst from the water, gasping for air. They clawed the masks from their faces and trod water in a circle around me and talked all at once. What was the ship carrying when it sank? Where was it going? What happened to the people?

We are earth-divers, and the earth is made of stories. I told the boys to dive and they would find them.

Notes

Chapter 1

[p. 1] The oceanic aspects of the Great Lakes have been commented upon by many authors, most famously by Herman Melville in *Moby-Dick* (New York: Everyman's Library, 1988 [1851]). Speaking of Steelkilt, "a Lakeman and desperado from Buffalo," Melville writes:

For in their interflowing aggregate, those grand freshwater seas of ours,—Erie, and Ontario, and Huron, and Superior, and Michigan,—possess an ocean-like expansiveness, with many of the ocean's noblest traits; with many of its rimmed varieties of races and of climes. They contain round archipelagoes of romantic isles, even as the Polynesian waters do; in large part, are shored by two great contrasting nations, as the Atlantic is; they furnish long maritime approaches to our numerous territorial colonies from the East, dotted all round their banks; here and there are frowned upon by batteries, and by the goat-like craggy guns of lofty Mackinaw; they have heard the fleet thunderings of naval victories; at intervals, they yield their beaches to wild barbarians, whose red painted faces flash from out their peltry wigwams; for leagues and leagues are flanked by ancient and unentered forests, where the gaunt pines stand like serried lines of kings in Gothic genealogies; those same woods harboring wild Afric beasts of prey, and silken creatures whose exported furs give robes to Tartar Emperors; they mirror the paved capitals of Buffalo and Cleveland, as well as Winnebago villages; they float alike the full-rigged merchant ship, the armed cruiser of the State; the steamer, and the beech canoe; they are swept by Borean and dismasting blasts as direful as any that lash the salted wave; they know what shipwrecks are, for out of sight of land, however inland, they have drowned full many a midnight ship with all its shrieking crew. Thus, gentlemen, though an inlander, Steelkilt was wild-ocean born, and wild-ocean nurtured; as much of an audacious mariner as any (p. 264).

[p. 2–3] The lake sturgeon (*Acipenser fulvescens*) is found throughout the Great Lakes drainage, as well as in the Hudson Bay and Mississippi drainages, and was once common (though now is approaching extinction) in the Ohio River drainage and in Alabama's Coosa River system. It is a bottom feeder that uses its protrusible mouth to suck up silt and mud and strain out food organisms. Some individuals up to nine feet long and 800 pounds have been reported in the Great Lakes, but seven feet long and 300 pounds in weight is the usual maximum size. Adults do not reach sexual maturity until they are twenty years old, according to Carl L. Hubbs and Karl F. Lagler, *Fishes of the Great Lakes Region* (Ann Arbor: University of Michigan Press, 1983). Some individuals have lived up to one hundred years of age.

Because the sturgeon spawns in rivers, it is vulnerable to pollution, poaching, and the loss of spawning areas to siltation and damming. The population has decreased steadily in the last century, no doubt due to river degradation and overharvesting. Although caviar made from the roe of the lake sturgeon is not as desirable as that from sturgeon species native to the Black and Caspian seas and the Danube River, the demand was great enough early in the twentieth century nearly to eradicate sturgeon from the Great Lakes. Their numbers remain low, but little else is known about their status.

[p. 3–4] The surface area of the Great Lakes and the length of their shorelines are rather variable numbers. The figures I quote are from *Coordinated Great Lakes Physical Data* (1992), compiled by the Coordinating Committee on Great Lakes Basic Hydraulic and Hydrologic Data. Shore length includes islands and the major channels between the lakes: the St. Mary's, St. Clair, Detroit, and Niagara rivers. See *The Great Lakes Basin*, Extension Bulletin E-1865, published by Michigan State University, in cooperation with Michigan Sea Grant, October 2000.

[pp. 5–6] Tocqueville's incomplete travels on the Great Lakes did not prevent him from making broad observations about them. Although true of only some portions of the lakes, his comments in *Democracy in America* (New York: Random House, 1945), remain interesting: "The Great Lakes which bound this region are not walled in, like most of those in the Old World, between hills and rocks. Their banks are flat and rise but a few feet above the level of their waters, each thus forming a vast bowl filled to the brim. The slightest change in the structure of the globe would cause their waters to rush either towards the Pole or to the tropical seas" (pp. 18–19).

[p. 8] Walter J. Hoagman, *Great Lakes Coastal Plants* (Lansing: Michigan State, University Board of Trustees, 1994).

[p. 10] The Emerson quote is from his essay "Nature."

Chapter 2

[p. 11] When discussing the color and clarity of water, it's helpful to differentiate between "apparent" color and "real" color. Much of the water in the Great Lakes is clear and thus has no "real" color. Real colors result from dissolved and suspended materials such as clay, minerals, and algae.

The apparent color of the Great Lakes is produced when light enters clear water and is absorbed and reflected by it. The water appears blue because the blue wavelengths of the visible light spectrum penetrate deeper than other wavelengths. The clearer the water, the deeper the penetration of blue wavelengths. Particles of sediment or microorganisms reduce penetration or "transmissivity" of light. Even relatively few particles can turn the "real" color of water to brown, green, red, or yellow. A good example is Wisconsin's Green Bay, where blooms of single-celled algae have historically tinted the water the color of chlorophyll.

Color is also affected by reflections of light on the surface. Sunlight directly above a body of smooth water is reflected only about 2 percent. When the sun is at an angle of 60 degrees, reflection increases to 6 percent. In both those cases, the water is likely to maintain its apparent blue color. At 90 degrees, however—when the sun rises or sets—the reflection of sunlight is nearly 100 percent, painting the surface of the water with the reds, oranges, and pinks of sunrise and sunset.

This subject is discussed at greater length in Jerry Dennis and Glenn Wolff, *The Bird in the Waterfall* (New York: HarperCollins, 1996).

[p. 13] For a more detailed description of the causes of "lake-effect" snow, see climatologist Val L. Eichenlaub's *Weather and Climate of the Great Lakes Region* (Notre Dame, IN: University of Notre Dame Press, 1979).

[p. 17] Much information on schooners, packet steamers, and other early vessels on the Great Lakes can be found in Walter Havighurst's *The Long Ships Passing* (New York: Macmillan, 1961).

[pp. 22–23] Seiches were frequently noted by early visitors to the Great Lakes. In 1689, Baron La Hontan noted tidelike variations of three feet per day in Green Bay. He also noticed the unusual, conflicting currents in the Straits of Mackinac. "The cause of this diversity of currents could never be fathomed," he wrote in *Voyages to Canada*; ". . . the decision of this matter must be left to the disciples of Copernicus"—Mentor L. Williams, ed., *Schoolcraft's Narrative Journal of Travels* (Lansing: Michigan State University Press, 1992), p. 245.

The French Jesuit Pierre de Charlevoix noted the same phenomenon on Lake Ontario in 1721, and wrote in his *Journal of a Voyage to North-America* (1761): "As I was walking on the banks of the lake I observed . . . that in this lake, and I am told that the same thing happens in all the rest, there is a sort of flux and reflux almost instantaneous, the rocks near the banks being covered with water, and uncovered again several times in the space of a quarter of an hour, even should the surface of the lake be very calm, with scarce a breath of wind. After reflecting for some time on this appearance, I imagined it was owing to the springs at the bottom of the lakes, and to the shock of their currents with those of the rivers, which fall into them from all sides, and thus produce those intermitting motions" (quoted in *Schoolcraft's Narrative Journal of Travels*, p. 245).

Henry Rowe Schoolcraft observed water levels carefully while canoeing around the lakes during the Cass expedition of 1820. After several days in Green Bay spent watching apparent tidal fluctuations and measuring them against a stake driven into the water's edge, Schoolcraft concluded:

. . . there are no regular tides in the lakes, at least, that they do not ebb and flow twice in twenty-four hours, like those of the ocean—that the oscillating motion of the waters is not attributable to planetary attraction—that it is very variable as to the periods of its flux and reflux, depending upon the levels of the several lakes, their length, depth, direction, and conformation—upon the prevalent winds and temperatures, and upon other extraneous causes, which are in some measure variable in their nature, and unsteady in their operation.

. . . These winds would almost incessantly operate, to drive the waters through the narrow strait of Michilimackinac, either into lake Huron or lake Michigan, until, by their natural tendency to an equilibrium, the waters thus pent, would re-act, after attaining a certain height, against the current of the most powerful winds, and thus keep up an alternate flux and reflux, which would always appear more sensibly in the extremities and bays of the two lakes; and with something like regularity, as to the periods of oscillation; the velocity of the water, however, being governed by the varying degrees of the force of the winds. (*Schoolcraft's Narrative Journal of Travels*, pp. 245–46.)

[p. 23] The details of the fatal 1929 and 1938 seiches on Lake Michigan are from witnesses' accounts published in the *Holland (MI) Sentinel*, August 20, 2001.

Chapter 3

[pp. 25–26] Nothing is solid, not even the ground beneath our feet. The earth flips magnetic poles, continents drift across the surface of the planet, oceans come and mountains go. All matter changes constantly, but often too slowly for comprehension. Most of us find it impossible to fathom the scale of human history, let alone geologic time. Only geologists, says John McPhee, trained to think in vast sweeps encompassing hundreds of millions of years, can "see the unbelievable swiftness with which one evolving species on the earth has learned to reach into the dirt of some tropical island and fling 747s into the sky"—*Basin and Range* (New York: Farrar, Straus & Giroux, 1980), p. 128.

In geologic time, rocks flow through a cycle analogous to the cycle that changes ocean to cloud to rain to river to ocean again. Molten rock rises to the surface, where it cools, hardens, and transforms into igneous rock. In the air, exposed to weather, the rock is chipped at by rain and ice, by heat and cold. Day after day, season after season, it crumbles and settles as sediment. Gravity compacts the sediment into sedimentary rock. Heat and pressure metamorphose the sedimentary rock into metamorphic rock. The subduction of tectonic plates drives the metamorphic rock deeper into the earth, where heat and pressure increase, and the rock becomes molten again, completing the cycle.

[pp. 26–27] The sinking of land beneath the weight of glaciers and ice sheets is referred to by geologists as "crustal subsidence." When the ice retreats, the land rises again in "isostatic rebound."

For more about the formation of the Great Lakes, see John A. Dorr, Jr., and Donald F. Eschman, *Geology of Michigan* (Ann Arbor: University of Michigan Press, 1971); Gene L. LaBerge, *Geology of the Lake Superior Region* (Tucson, AZ: Geoscience Press, 1994); and Thomas F. Waters, *The Superior North Shore* (Minneapolis: University of Minnesota Press, 1999).

[pp. 27–28] Green Bay was known by the early French as *Baye de Puants*, "Bay of Stinkers." The historian David Lavender, like others before him, assumed the name referred to the Winnebago people who lived along the bay. The Winnebagos, he wrote in *The Fist in the Wilderness* (Garden City, NY: Doubleday, 1964), were "the most sullen and unlovable of the Northwestern tribes," and subsisted on a diet that caused "so overwhelming a stench that the French called the tribe *Puants*, or Stinkers, and their fishing grounds *La Baye de Puants*. Later the foliage on the shores produced a more euphonious name, *La Baye Verte*, or, today, Green Bay" (p. 6).

More recent scholars have argued that "Bay of Stinkers" was a corruption of the Winnebagos' tribal name. In their own language, *Winnebago* means "stinking water," in reference to the Fox River, which enters the bay after draining prairies and swamps, where it presumably became rank from decaying vegetation.

William Ashworth notes in *The Late Great Lakes: An Environmental History* (Detroit: Wayne State University Press, 1986) that the fertile Fox River contributed to Green Bay's natural eutrophication, a condition that in modern times has been greatly amplified by industrial and municipal wastes. Clustered within the Green Bay drainage is one of the highest concentrations of pulp and paper mills in North America. Their effluent includes large amounts of ammonia, a byproduct of ammonium bisulfate, which is used as a solvent to free cellulose fibers from woodpulp. Ammonia can be toxic, but it is also a nitrogen compound and a nutrient that promotes plant growth. In Ashworth's words, it makes a "nice little positive feedback loop. . . . The resulting overgrowth of algae raises the surrounding water's pH level, making it less acidic: The higher pH levels slow down the decomposition of the ammonia, making more of it available for use as plant food. The circle goes on and on, the snake eating its own tail forever and ever, each new increment of ammonia added to the water causing more eutrophication than the last" (p. 138).

[p. 29] The two Grand Traverse bays, unlike most large bays in the Great Lakes, are deep, clear, and cold, and therefore support the same aquatic organisms found in the open waters of the lake. In fact, biologists consider West Bay a microcosm of Lake Michigan itself. The bay has approximately the same shape as the big lake, is oriented north and south, and has a metropolitan center at its base. A recent biological invader, the tiny zooplankton *Cercopagis pengoi*, or fishhook flea, was discovered in West Bay in autumn 1999, months before it was found elsewhere in Lake Michigan. Other biological indicators suggest big changes to come. Zebra mussels have colonized both bays and seem to have already led to increased water clarity. In the past couple of years the population of rusty crayfish has exploded. These alien crawdads hide beneath every rock; when you wade through the shallows they dart ahead like an infestation of mice in a field.

[p. 32] Ernest Hemingway, *The Nick Adams Stories* (New York: Scribner Paperback, 1999), p. 239.

[pp. 34–35] The history of James Jesse Strang and his "kingdom" on Beaver Island is told in Roger Van Noord's *The Assassination of a Michigan King: The Life of James Jesse Strang* (Ann Arbor: University of Michigan Press, 1997).

[pp. 39–40] The buoy lights we passed at Gray's Reef were not lit arbitrarily, nor did they contradict the standard green-on-starboard, red-on-port navigation rules of boats and aircraft. Buoy lights and channel markers are arranged according to a universal alliterative law: "Red Right Return." This means that a boat returning to harbor from the ocean can always expect to be guided by red lights and other markers on the right (starboard) side, and green on the left (port) side. Marine lawmakers long ago decided that the Great Lakes should be considered one large harbor—thus a boat entering them and traveling upbound from the ocean is "returning" and should keep all red buoys and markers on the right side. Since the *Malabar* was downbound, headed for the ocean, we reversed the law of "Red Right Return" and kept green on our right, red on our left.

Chapter 4

[pp. 41–42] The full story of Chicago's rise from the ashes and the creation of its spectacular waterfront is told in Donald L. Miller's *City of the Century: The Epic of Chicago and the Making of America* (New York: Simon & Schuster, 1996).

[pp. 49–50] In a race employing the handicap system, the boat with the fastest time does not necessarily win. In 1987, a 68-footer named *Pied Piper* finished the Chicago-to-Mackinac Race in a little under 26 hours, setting an elapsed-time record that stood until Steve Fossett broke it a decade later. Yet *Pied Piper* was not the divisional winner. After correcting for handicap, the award was given to a vessel that crossed the finish line behind *Pied Piper* and nine other faster boats.

[pp. 50–51] Seasickness has a variety of symptoms, but none so disagreeable as vomiting. My knowledge of the physical act of emesis was increased more than I needed to know by an article in *The New Yorker* (July 5, 1999) by Dr. Atul Gawande. According to Dr. Gawande, when my stomach lurched for the first time and I felt blood drain from my skin as my capillaries constricted (causing me to feel chilled and probably turn pale), I was entering the "prodromal phase of emesis." Even with cold wind in my face, I perspired—the notorious cold sweats. Saliva filled my mouth. My heartbeat increased. My pupils dilated. My stomach convulsed, as if wrung by strong hands, and my esophagus clenched, lifting my upper stomach into my chest. Next came "the retrograde giant contraction," in which the upper part of the small intestine contracted rhythmically, causing it to pump yesterday's food back into the stomach (which is why I vomited more than just breakfast). Next came retching, caused by contractions of the muscles in the abdomen, diaphragm, and chest, followed quickly by expulsion: a single prolonged contraction of the abdomen and diaphragm, which pressurized the stomach. When the esophagus relaxed, the pressure was released. In the memorable words of Dr. Gawande, it was as if "someone had taken the plug off a fire hydrant."

[p. 53] For a vivid account of the storm that struck the Irish Sea in 1979, see John Rousmaniere, *Fastnet Force 10* (New York: W. W. Norton, 1980).

[p. 57] Henry Schoolcraft's description of Sleeping Bear Dunes is from Williams, ed., *Schoolcraft's Narrative Journal of Travels,* p. 262. Pierre Charlevoix's words are from his *Journal of a Voyage to North-America* (quoted in Schoolcraft, p. 262).

[pp. 57–62] Great Lakes dunes are discussed in detail in Edna and C. J. Elfont, *Sand Dunes of the Great Lakes* (Chelsea, MI: Sleeping Bear Press, 1997). See also the booklet *Discovering Great Lakes Dunes,* produced by Michigan State University Extension and Michigan Sea Grant. Much of my knowledge of sand mining comes from *Vanishing Lake Michigan Sand Dunes: Threats from Mining,* a report produced by the Lake Michigan Federation, a Chicago-based environmental group.

[pp. 59–60] The processes of sand on a beach are explained in detail in Dorr and Eschman, *Geology of Michigan.*

[p. 63] The quote by Francis Comte de Castlenau is from George Weeks, *Sleeping Bear: Yesterday and Today* (Franklin, MI: A&M Publishers, 1990), pp. 124–126.

Chapter 5

[p. 69] Another "mackinaw" is worth noting, this one familiar far beyond the Great Lakes. In the winter of 1811, a garrison of British soldiers was stationed on St. Joseph Island, a large island in the St. Mary's River a short distance north of the Straits. The winter was harsh and the troops were underdressed and miserable. Because no supplies were scheduled to arrive until spring, the commander of the garrison was forced to requisition woolen Hudson's Bay blankets from a local trader and put the women of the island to work cutting and sewing them into garments. The short, belted, colorful wool jackets they produced are manufactured to this day under the name "mackinaw coats."

[pp. 69–70] After decades of consultation, feasibility studies, bond sales, and legislative debates, work on the Mackinac Bridge began on May 7, 1954. The first step was to fill enormous caissons with rock and cement and lower them until they were seated in the bedrock bottom of the Straits, where they would support the twin forty-six-story-high towers at the middle of the bridge. For the full story of the bridge's construction, see Lawrence A. Rubin, *Mighty Mac: The Official Picture History of the Mackinac Bridge* (Detroit: Wayne State University Press, 1989).

Currently the Mackinac Bridge is the third longest suspension bridge in the world. The longest, built in 1998, is Japan's Akashi Kaikyo Bridge, with a total suspension of 12,826 feet. The second longest, also completed in 1998, is Halsskov-Sprogoe in Denmark, at 8,921 feet. The Mackinac remains the longest suspension bridge in the western hemisphere.

[p. 73] The Father Marquette quote is from Clifton Johnson, *Highways and Byways of the Great Lakes* (New York: Macmillan, 1911), p. 154.

[pp. 74–75] For more about the history of the settlements and forts in the Straits and on Mackinac Island, see Walter Havighurst's *Three Flags at the Straits* (Englewood Cliffs, NJ: Prentice-Hall, 1966).

[pp. 77–78] For a more complete history of Straits shipwrecks (including much information useful to divers), see Charles E. and Jeri Baron Feltner, *Shipwrecks of the Straits of Mackinac* (Dearborn, MI: Seajay Publications, 1991).

Chapter 6

[p. 81] I have written in greater detail about the blizzard of 1978 in an essay, "A Good Winter Storm," collected in my book, *The River Home* (New York: St. Martin's Press, 1998). Since publication of that book, many readers have written to share their own memories of the storm, which paralyzed much of the middle and eastern United States and is considered one of the most severe winter storms of the century.

[p. 86] Lichen and other emergency edibles have often found their way into soup pots. In *La Salle and the Discovery of the Great West* (reprinted New York: Modern Library, 1999), Francis Parkman reports that Jesuit missionary Louis André spent the winter of 1670 on the shore of Lake Huron, where food was so scarce that the "staple of his diet was acorns and *tripe de roche,* a species of lichen, which, being boiled, resolved itself into a black glue, nauseous, but not void of nourishment. At times he was reduced to moss, the bark of trees, or mocassins and old moose-skins cut into strips and boiled" (pp. 25–26). The voyageurs ate somewhat better as a rule, though they were occasionally forced to boil mocassins when they happened to become stranded by early winter storms. For most meals, the voyageurs while traveling from Montreal and around the lakes ate a gruel of peas, corn mush, and water mixed with pork fat—leading to the derogatory name "Pork Eaters." Inland and west of Lake Superior, they were more likely to eat pemmican made of dried and shredded bison meat mixed with buffalo lard and spiced with dried berries. This could be eaten boiled or raw and would last for months without refrigeration. Because pemmican was so high in calories, the voyageurs could get by on two meals a day. Sometimes they varied their menu by adding water, flour, and maple syrup to their pemmican, making it into a soup called "rubbaboo."

[p. 86] The translation of *Souffle, souffle, la Vieille*—"Blow gently, blow gently, Old Woman"—is from Marjorie Cahn Brazer, *The Sweet Water Sea* (Manchester, MI: Heron Books, 1984), p. 173.

Early attitudes about the lake can be viewed a few miles south of the Michipicoten River, where ancient Ojibwa pictographs adorn granite cliffs along the shore of Lake Superior Provincial Park. Painted on the rock in red ocher are representations of sturgeon, moose, birds, dancing figures, and a canoe with two paddlers. One large pictograph represents a daring crossing of Lake Superior by fifty men in canoes. The crossing, from the Porcupine Mountains in Michigan's western Upper Peninsula, was blessed by the deity Misshepezhieu, represented as a lynxlike being with prominent horns. The pictographs can be reached down a half-mile path and along a rather precarious ledge so close to the water's edge that anything less than calm water makes walking treacherous to impossible. My luck has been poor. In three attempts I've never been able to negotiate the ledge.

[p. 87] The Ojibwa legend about coming to the Great Lakes in search of "food on the water" is found in Ron Morton and Carl Gawby, *Talking Rocks: Geology and 10,000 Years of Native American Tradition in the Lake Superior Region* (Duluth: Pfeifer-Hamilton Publishers, 2000).

[p. 89] The poet Keith Taylor, who grew up in British Columbia in the 1950s and 1960s and now teaches writing and literature at the University of Michigan, tells me that Canadian schoolchildren of his generation referred to the voyageurs Radisson and Groseilliers as "Radishes and Gooseberries." The kids knew their French. The Pigeon River, on the Superior coast of Minnesota, was labeled the Groseilliers River on early maps, but an English cartographer appropriated the name and bestowed its translation upon another nearby stream: the Gooseberry.

An early English translation of Radisson's narrative was discovered at the end of the nineteenth century. It gives a vivid and eccentrically spelled description of Superior, a lake "most delightfull and wounderous," with the Pictured Rocks on the Michigan shore rising "like a great Portall, by reason of the beating of the waves." In that document Radisson summarized the feelings of many early white explorers and traders who enjoyed freedom from authority in the Old Northwest: "We weare Cesars being nobody to contradict us."

[pp. 90–91] Enormous profits were made in the fur trade, but not by those who did the actual labor. John Jacob Astor and a handful of other magnates realized the greatest profits, amassing legendary fortunes (Astor went on to invest much of his in Manhattan real estate). Even merchants further down the heirarchy were in position to grow wealthy. In 1836, Washington Irving published *Astoria*, a history of his friend John Jacob Astor's expansion of the fur trade from the Old Northwest to Oregon. Irving did not visit the Great Lakes, and his book often reads like a press release designed to bolster Astor's image, but the history is largely accurate. Irving itemizes the profits made about 1810 by licensed Montreal merchants who employed *coureurs de bois* to trade with Indian trappers in the interior. Setting off in the spring in canoes loaded with trade goods, the *coureurs de bois* would return before freeze-up the next winter with their canoes filled with baled furs, which the merchants would sell to exporters who delivered them to markets in the major cities of Europe and the eastern United States. He writes:

> The following are the terms on which these expeditions were commonly undertaken. The merchant holding the license would fit out the two canoes with a thousand crowns' worth of goods, and put them under the conduct of six coureurs des bois [sic], to whom the goods were charged at the rate of fifteen per cent above the ready money price in the colony. The coureurs des bois, in their turn, dealt so sharply with the savages that they generally returned, at the end of a year or so, with four canoes well laden, so as to insure a clear profit of seven hundred per cent, insomuch that the thousand crowns invested, produced eight thousand. Of this extravagant profit the merchant had the lion's share. In the first place he would set aside six hundred crowns for the cost of his license, then a thousand crowns for the cost of the original merchandise. This would leave six thousand four hundred crowns, from which he would take forty per cent, for bottomry, amounting to two thousand five hundred and sixty crowns. The residue would be equally divided among the six wood rangers, who would thus receive little more than six hundred crowns for all their toils and perils.

[p. 94] The Charlevoix quote is from his *Histoire de la Nouvelle France*, first published in 1744 and translated into English in 1900. His journal of the trip, composed of a series of letters to the duchess de Lesdiguières, was published in 1761 as *Journal of a Voyage to North-America*.

[p. 96] The legend of the snow wasset is collected in Victoria Brehm's *Sweetwater, Storms, and Spirits: Stories of the Great Lakes* (Ann Arbor: University of Michigan Press, 1991), pp. 122–24.

[pp. 97–98] For more about the natural history of the Lake Superior region, see Thomas F. Waters, *The Superior North Shore*.

Chapter 7

[p. 101] The average temperature differential for November 10 is from a *Detroit News* story, "The Gales of November," November 2, 1989. Other information on Great Lakes weather patterns as they relate to the storms of November is from Eichenlaub, *Weather and Climate of the Great Lakes Region*.

[pp. 101–106] Many books have been written about shipwrecks in the Great Lakes. I'm indebted in particular to two by Robert J. Hemming, *Ships Gone Missing: The Great Lakes Storm of 1913* (Chicago: Contemporary Books, 1992), and *The Gales of November: The Sinking of the Edmund Fitzgerald* (Chicago: Contemporary Books, 1981). Other informative books include Walter Havighurst's *The Long Ships Passing* (New York: Macmillan, 1961); William Ratigan's *Great Lakes Shipwrecks and Survivals* (Grand Rapids, MI: William B. Eerdmans, 1960); and Frank Barcus's *Freshwater Fury* (Detroit: Wayne State University Press, 1960). For more about the *Fitzgerald*, see Frederick Stonehouse, *The Wreck of the Edmund Fitzgerald* (AuTrain, MI: Avery Color Studios, 1977), and Hugh E. Bishop, *The Night the Fitz Went Down* (Duluth: Lake Superior Port Cities Inc., 2001).

[p. 105] The year after the sinking of the *Fitzgerald*, the Coast Guard submerged remote cameras 500 feet into Superior and located the wreck. Jacques Cousteau's *Calypso* arrived in 1980, and initiated the first manned dive, in a submersible. None of those expeditions, nor several others by divers in deep-sea equipment (and one in scuba gear, breathing a special mixture of gases that allow extremely deep dives), found conclusive evidence to determine the cause of the sinking.

[p. 105] In the years since the sinking of the *Edmund Fitzgerald*, and in part because of the loss of that ship, commercial vessels on the lakes have become safer for their crews and passengers. All ships are now equipped with global positioning systems (GPS) that use satellites to navigate and pinpoint their locations, making it considerably easier and quicker to locate vessels in emergencies. Vessels are also equipped with sonar depth finders (the *Fitzgerald* was not, though the technology was available), and crewmen are issued cold-water survival suits outfitted with radio position beacons.

The myth endures that Lake Superior never gives up her dead. Gordon Lightfoot perpetuated the notion in his ballad about the *Edmund Fitzgerald*, but it goes back much further. In an article entitled "The Great Lakes" (originally copyrighted 1893; reprinted 1976 in *Stories of the Great Lakes* [Grand Rapids, MI: Black Letter Press]), W. S. Harwood wrote: "Whoever encounters terrible disaster . . . and goes down in the angry, beautiful blue waters [of Lake Superior], never comes up again. From those earliest days when the daring French voyageurs in their trim birch-bark canoes skirted the picturesque lake, down to this present moment, those who have met their deaths in mid-Superior still lie at the stone-paved bottom. It may be said that, so very cold is the water, some of their bodies may have been preserved through the centuries" (p. 18).

Chapter 8

[p. 121] For more about the geology and history of Georgian Bay, see Brazer's *The Sweet Water Sea* and James Barry's *Georgian Bay: The Sixth Great Lakes* (Erin, Ontario: Boston Mills Press, 1995). An interesting geological study is William Gillard and Thomas Tooke, *The Niagara Escarpment* (Toronto: University of Toronto Press, 1978).

[pp. 125–126] Even the earliest Europeans recognized the value of Michigan's natural resources and made efforts to purchase the land from its original owners. Shortly after the War of 1812, the Michigan Territory was established as a network of trading posts, each on a major river. In 1819, Lewis Cass, who would later become the state's first governor, persuaded the Chippewas to give up their claim to the northeast portion of the Lower Peninsula, from Saginaw Bay to Thunder Bay. The remainder of the Lower Peninsula and about half of the Upper Peninsula were procured by 1836, one year before Michigan's official statehood. By 1842, virtually all Indian lands had been acquired by the state.

[pp. 126–127] Some of the statistics and other details about the lumber industry are from James Kates, *Planning a Wilderness: Regenerating the Great Lakes Cutover Region* (Minneapolis: University of Minnesota Press, 2001). See also Dave Dempsey's insightful and meticulous history of Michigan's environmental legacy, *Ruin and Recovery: Michigan's Rise as a Conservation Leader* (Ann Arbor: University of Michigan Press, 2001).

[pp. 128–131] Details of the Peshtigo fire are from Stewart H. Holbrook's *Burning an Empire: The Story of American Forest Fires* (New York: Macmillan, 1952), and from Jerry Resler, "Where the World Ended: Peshtigo marks 125th anniversary of fire that killed 1,200," *Milwaukee Journal Sentinel*, September 15, 1996.

The Michigan fires are described in Holbrook and in Betty Sodder, *Michigan on Fire* (Lansing, MI: Thunder Bay Press, 1997).

[p. 132] Dioxins and other persistent bioaccumulative toxins entered Saginaw Bay from Dow Chemical Company plants in Midland, which for decades dumped its wastes directly into the Tittabawassee River, a tributary of the Saginaw River. For the story of Dow's resistance to environmental controls (and ultimate compliance), see Dempsey's *Ruin and Recovery: Michigan's Rise as a Conservation Leader*.

Chapter 9

[p. 145] The information about Harsens Island and the St. Clair Flats is from Tom Carney's *Natural Wonders of Michigan: A Guide to Parks, Preserves, and Wild Places* (Castine, ME: Country Roads Press, 1995).

Some of the difficulties of navigating on the St. Clair Flats are described in *Thompson's Coast Pilot* for the upper Great Lakes, 1869 edition. The author notes with cheer: "Since the last edition was printed I am happy to state that there are a great many improvements going on, such as opening new cuts for channels, building piers, lighthouses, beacons, placing ranges for harbors and laying down buoys, etc., many of which are finished, and others to be pushed forward as fast as possible.

"The first improvement in point of utility is the St. Clair Flats, where a new straight cut is being made, of a depth of water sufficient for the largest class of vessels. This cut or channel, when finished, will be of the greatest importance to the merchant, as well as to the sailing community."

[pp. 145–149] The story of the *Griffin* is from Francis Parkman's 1869 history, *La Salle and the Discovery of the Great West*. See also Russell McKee, *Great Lakes Country* (New York: Thomas Y. Crowell Co., 1966), and William Ratigan's *Great Lakes Shipwrecks and Survivals*.

Chapter 10

[p. 159] Although the blue pike, a close relative of the walleye, is officially considered extinct, rumors have circulated for years that remnant populations exist in certain inland lakes in Ontario. To determine if that is so, scientists need to compare the DNA of the living fish with that of actual blue pike. But such testing requires tissue samples, and no blue pike tissues are known to exist. Hope was raised when the *New York Times* (March 18, 1999) reported the discovery of a blue pike that a Conneaut, Ohio, barber had kept preserved in his home freezer since 1962. If DNA analysis of the frozen specimen matched the blue-tinted fish of the inland lakes, it would prove that the blue pike was not extinct after all and perhaps could be reintroduced to Erie. Unfortunately, the frozen specimen turned out to be a walleye.

[pp. 166–167] The quote by Tom Nalepa is from a December 4, 1997, NOAA press release, "Environmentally Sensitive Organisms Missing in Lake Michigan Mud Samples."

[p. 167–169] A detailed history of the worst polluters of the Great Lakes is found in Ashworth, *The Late Great Lakes*, p. 155.

[p. 169] A report by the EPA in early 2002 warns that about 20 percent—some 2,000 miles—of Great Lakes shoreline is polluted with toxic substances and industrial byproducts, most of it contained in bottom sediment. The 1978 Great Lakes Water Quality Agreement defined a "toxic substance" as one that "can cause death, disease, behavioral abnormalities, cancer, genetic mutations, physiological or reproductive malfunctions or physical deformities in any organism or its offspring, or which can become poisonous after concentration in the food chain or in combination with other substances." It is considered a "persistent toxic substance" if it has a half-life in water greater than eight weeks ("half-life" is the time required for half the amount of substance to be eliminated naturally from a system). These definitions and the standards established by the Great Lakes Water Quality Agreement are discussed in more detail in *Great Lakes, Great Legacy?* by T. Colborn, A. Davidson, S. Green, et al. (Washington, DC: Conservation Foundation, 1990). Chemical contamination of the Michigan waters of the Great Lakes is detailed in Dempsey's *Ruin and Recovery: Michigan's Rise as a Conservation Leader.*

Listed below are the most critical pollutants in the Great Lakes, according to the International Joint Commission, with elaboration from the *Encyclopedia of the Environment,* ed. Ruth Eblen and William Eblen (Boston: Houghton Mifflin, 1994):

—Polychlorinated biphenyls (PCBs): This group of 209 related chemicals includes fluids used as insulation in eletrical transformers and in the production of inks, lubricants, and hydraulic fluids. Medical researchers suspect that even small doses of PCBs can cause liver damage and developmental problems in children, but lab tests to determine whether they are carcinogens have been inconclusive. Although their use is now banned, they continue to enter the lakes through atmospheric deposition and in sediments.

—2,3,7,8,-tetrachlorodibenzo-p-dioxin (TCDD): This is the most toxic of the seventy-five forms of dioxins, and was an ingredient in the infamous Agent Orange, a defoliant used in the Vietnam War, but has been used as well in agriculture and forestry herbicides in North America. This and other dioxins are also byproducts of burning fossil fuel and wastes, and of the processes used to produce paper and paper pulp.

—2,3,7,8-tetrachlorodibenzofuran (TCDF): This most toxic of the 135 types of furan is used in herbicides for agriculture and forest management, and is a byproduct of pulp and paper production and the burning of fossil fuel.

—Dichlorodiphenyltrichloroethane (DDT): Introduced in the 1930s and once the most widely used insecticide in the world, DDT has been banned in the United States since 1972. It is still widely used elsewhere in the world for mosquito control, especially in tropical regions, and continues to enter North American waters via atmospheric deposition and sediment disruption. Although no conclusive evidence has been found to prove that DDT is harmful to human health, it produces thin and weakened eggshells in birds of prey and decimated populations of osprey, eagles, and others before it was banned in the United States.

—Dieldrin: This insecticide is sprayed on fruit trees and enters the Great Lakes via air and sediments.

—Toxaphene: An insecticide developed to replace DDT and used extensively in cotton production. It enters the Great Lakes from air and sediments.

—Mirex: Used both as a fire retardant and a pesticide, especially to control fire ants. When exposed to sunlight, it breaks down into a more potent chemical, photomirex. It is found in sediments and enters water through the atmosphere.

—Mercury: This liquid metallic element is used in paints, paper manufacture, electrical switches, thermostats, dry-cell batteries, and fluorescent lights, and is a by-product of many industrial processes. It occurs naturally in soils and sediments. Like most heavy metals, it is highly toxic; it is also volatile at low or moderate temperatures, which results in mercury vapor being released into the atmosphere when products containing the element are burned. The vapor can travel long distances in the air, then settles to the ground with precipitation. High exposure can cause nervous system disorders, personality disturbances, blindness, deafness, and death.

—Alkylated lead: This heavy metal is used for bullets, batteries, and fishing sinkers, as a fuel additive, and in paints, solder, and pipes. It is released into the atmosphere when fossil fuels and waste products are burned. Symptoms of lead poisoning in humans include high blood pressure, infertility, hearing loss, and decreased intelligence.

—Benzopyrene: A polyaromatic hydrocarbon released by internal combustion engines and by the burning of fossil fuels, wood, charcoal, and various wastes.

—Hexachlorobenzene (HCB): A byproduct given off when fossil fuels and wastes are burned and chlorine is manufactured.

[pp. 170–171] Some contaminated sites around the Great Lakes qualify for funding from the EPA's Superfund; others must be paid for by the polluting industries or by individual states and Canada. The Superfund was established shortly after the Love Canal incident carried the issue of toxic contamination to public awareness. The Hooker Chemical Company spent years burying 21,800 tons of chemical waste in Love Canal, near Niagara Falls, New York. In 1953, they covered the site and sold it to the Niagara Falls Board of Education for one dollar. A suburban neighborhood sprang up, but homeowners soon noticed unpleasant tastes and odors in their drinking water and strange, noxious liquids seeping into their basements. In 1978, the neighborhood was declared a federal disaster area and

239 families were evacuated and ultimately relocated. In response, the U.S. Congress earmarked $1.6 billion to create a Superfund to pay for cleanup of toxic waste sites around the country. Today, only the most severely contaminated sites qualify for Superfund status.

[p. 172] For more about the bird life of Point Pelee and elsewhere in the lakes' region, see Tom Powers's *Great Birding in the Great Lakes* (Flint, MI: Walloon Press, 1998).

[pp. 172–174] Sudden squalls on Erie and the other lakes have taken on nearly mythological significance. Dwight Boyer, in his *Strange Adventures of the Great Lakes* (New York: Dodd, Mead & Co. 1974), describes a squall that struck Lake Erie on August 10, 1971, with winds up to 100 miles per hour. It was even more short-lived than the two that struck us on the *Malabar*—it lasted only seven minutes. Boyer proposes that such "unpredictable and devilishly capricious freaks of weather" are "white squalls":

> As yet unexplained by meteorological science, they are stealthy, sudden and violent—whirlwind-like tempests which seem to spring from nowhere to maul and buffet their victims, briefly but frequently fatally. There is no visual clue or discernible evidence such as darkening skies, mounting seas or lowering clouds. Obviously then, there is no defense against the unforeseen. They strike day or night, under ideal sailing conditions, furiously twisting manifestations that some of nature's secrets have yet to be probed or revealed . . . undoubtedly many vessels, from the schooners of the Great Lakes to the big tea clippers of the deep seas, have "gone missing" from unknown causes when, in all truth, the probability was that some were overtaken and overwhelmed by that unseen scourge, the white squall. . . . A laden Great Lakes schooner with her canvas up would likely be shorn of her sails, shrouds, stays, and masts, possibly blown over on her beam ends to lie there helpless in the trough, prey for whatever winds and seas later developed. But the schooner devoid of cargo was the most vulnerable victim. Completely without bulk and weight low in her holds to counteract the skulking but invisible enemy, she would likely as not be capsized immediately. In the vernacular of those who survived such catastrophic maulings, the experience was described as being "knocked-down."

Meteorologists are more inclined to blame such sudden, violent winds on "downbursts" and "wind shear"—unusual but explainable phenomena caused when masses of air of varied temperatures collide in the atmosphere.

Chapter 11

[pp. 176–177] Dave Stone's books are *Long Point: Last Port of Call* (Toronto: Boston Mills Press, 1993) and *Waters of Repose: The Lake Erie Quadrangle*, with David Frew (Erie, PA: Erie County Historical Society, 1993).

[pp. 179–180] The anecdote about Long Point residents who salvaged whiskey bottles washing up on shore reminds me of a story closer to home. Historian Steve Harold reports that on November 19, 1879, the schooner *W. B. Phelps*, carrying a cargo that included a large quantity of beer, ran aground in a storm near Glen Arbor, in Lake Michigan's Manitou Passage. The captain and four crewmen were swept to their deaths by high waves, but two survivors managed to cling to the bow of the wreck. A call went out for volunteers to launch a rescue boat from shore, but the call went largely unheeded, "due to the size of the waves and to the fact that everyone was collecting beer as rapidly as possible." Eventually a crew was assembled, a boat was launched, and the two men were rescued. See Steve Harold's *Shipwrecks of the Sleeping Bear* (Traverse City, MI: Pioneer Study Center, 1984). A more detailed account of the rescue is found in George Weeks's *Sleeping Bear: Yesterday and Today*.

[p. 182] In fact, Niagara Falls is comprised of two waterfalls of uneven height. The American Falls are 167 feet high; Horseshoe Falls 158 feet.

[pp. 186–187] The story of the *Sanderson* is from Harold's *Shipwrecks of the Sleeping Bear*.

[pp. 189–190] The alewife, *Pomolobus pseudoharengus*, has been widespread throughout the Great Lakes since at least 1954, according to Hubbs and Lagler in *Fishes of the Great Lakes Region*. It was originally landlocked in Lake Ontario and some inland lakes connected by tributaries to the larger lake, and did not gain access to the upper lakes until the opening of the Welland Canal. In its saltwater habitat, which includes most of the Atlantic coast of North America from Labrador to Florida, the alewife is anadromous, spending most of its life in the ocean, ascending freshwater rivers and lakes at the age of three or four years to spawn. This is the fish so eloquently described in John Hay's nature classic, *The Run* (New York: Ballantine Books, 1971 [1959]). It is also the fish that Squanto taught the Pilgrims to use as fertilizer for their corn during their first April in Plymouth. In the centuries that followed, New Englanders netted huge quantities of alewives from their spawning streams and used them as fertilizer, cat and dog food, fish meal, and lobster bait. Cape Cod seafarers salted them, dried them in the sun, smoked them, and strung them on sticks to be eaten like corndogs. But the fish are too full of bones to be much enjoyed and their flavor caused one early chronicler, Marshal McDonald Douglas, to write in 1740 that "They are a very mean, dry and insipid fish. Some of them are cured in the manner of white Herrings and sent to the sugar islands for the slaves, but because of their bad quality they are not in request. . . ."—Quoted in *The Run*, p. 20.

They are not much in request in the Great Lakes, either, where they spend their entire lives in freshwater. Alewives in the lakes have the unfortunate habit of periodically dying off in large numbers. Nobody knows why this occurs, though biologists speculate that it might be related to the stresses of spawning or to fluctuations in available forage.

[pp. 189–190] Although TFM has proven to be an effective tool for controlling lampreys, is nontoxic to humans, and has met all EPA criteria for application in tributaries of the Great Lakes, it is expensive. Other techniques used to control sea lamprey include electrical and mechanical barriers and the experimental release of sterile males into the lamprey population. The four main types of barriers currently in use are low-head barriers (each with a two- to four-foot drop, often with a lip to prevent lampreys from climbing over with their suction-cup mouths), adjustable-crest barriers that use air bladders inflated only during spawning runs of lampreys, velocity barriers that create areas of very fast water over surfaces lampreys cannot attach to (the poor-swimming lampreys can't negotiate the current, but fish can), and electrical barriers which use direct current on the streambed, only during the spawning runs, to deter lampreys without affecting fish and other wildlife. Research underway includes an attempt to lure adult lampreys into traps with pheromones.

Many fisheries experts consider controlling the sea lamprey the most important work they can do in the Great Lakes. As history has amply demonstrated, when the lamprey population increases, the gamefish population decreases. "Left uncontrolled, even for a short time," says Charles Krueger, vice chair of the Great Lakes Fishery Commission, "lampreys will destroy the fishery."

[p. 190] One species of Pacific salmon had already established a self-sufficient population by the time coho and Chinook were planted in the late 1960s. In 1955, eggs of pink salmon—also called humpbacked salmon, for the spawning male's deformed spine—were flown from British Columbia to Thunder Bay, Ontario, near the shore of Lake Superior, with the intention of raising them in a hatchery and stocking them far to the north in Hudson Bay. When the hatched fingerlings were loaded onto seaplanes and transported north, however, about twenty thousand of them were inadvertently left behind at the hatchery. Rather than allow them to die, hatchery employees released them into the Current River, a tributary of Lake Superior, and they were quickly forgotten.

The attempt to introduce the salmon in Hudson Bay failed, but a few years later adult pink salmon began showing up in rivers around Lake Superior's Thunder Bay. Within ten years they were spawning in the St. Mary's River at the outlet of Lake Superior, and in tributaries along the north shore of Lake Huron. In the years since, they've established themselves in all of the Great Lakes and are found in good numbers in such unlikely places as the St. Clair River north of Detroit. But because of their small size—they average two or three pounds at maturity—pinks have never attracted much attention from either commercial or recreational anglers. One exception is in the rapids of the St. Mary's River, where heavy runs of pink salmon in August and September are popular with anglers using fly rods and other light tackle.

Chapter 13

[p. 210] The walking tours of John Bartram and Alexander Wilson are described in Joseph Kastner's *A Species of Eternity* (New York: Knopf, 1977).

Chapter 14

[pp. 219–222] For a more detailed history of the Erie Canal, see Carol Sheriff's scholarly study, *The Artificial River: The Erie Canal and the Paradox of Progress, 1817–1862* (New York: Hill & Wang, 1996).

[p. 221] A short biography of Samuel Latham Mitchell, and a portion of the speech he gave at the Erie Canal celebration in New York Harbor, is found in the chapter entitled "A Chaos of Knowledge" in Kastner's *A Species of Eternity*. According to Kastner, Mitchell was a polymath of amazing breadth and reach: "an anthropologist, botanist, chemist, entomologist, geologist, herpetologist, ichthyologist, mineralogist, ornithologist, paleontologist and zoologist" (pp. 194–195). He also served at various times in the New York State assembly, the house of representatives, and the senate.

[p. 233] Many of the toxic chemicals found in bottom sediments and in the fatty flesh of organisms in the Great Lakes, Hudson River, and countless other waterways are known to be carcinogenic or have been demonstrated to cause birth defects. PCBs are a special case, however. The 1979 decision by the EPA to prohibit manufacture of the chemicals in the United States and to phase out their use over the next five years was based on concerns that were later shown to be invalid. Specifically, the decision was made after fifteen hundred people in the Japanese city of Yusho became ill in 1975 when they ate rice that was accidentally contaminated with PCBs. Although the Centers for Disease Control and Prevention concluded in 1975 that PCBs caused liver cancer in rats, a decade later those findings were questioned by the same scientist who had conducted the original research; subsequent research found no evidence that exposure to PCBs caused cancer in humans. The EPA now considers the compounds to be "suspected carcinogens," reason enough, of course, for alarm.

For more on the environmental and natural history of the Hudson River, see John Cronin and Robert F. Kennedy, *The Riverkeepers* (New York: Scribners, 1997), and Robert H. Boyle, *The Hudson River: A Natural and Unnatural History* (New York: W. W. Norton, 1979).

Chapter 15

[p. 251] The "scrotumtightening sea" is mentioned by Buck Mulligan in the first section of *Ulysses*, where he expounds to Stephen Dedalus on the glories of the ancient Greeks and their "great sweet mother," the sea. Among the many pleasures of reading Joyce are his observations of the sea: "Inshore and farther out the mirror of water whitened, spurned by lightshod hurrying feet" . . . "Warm sunshine merrying the sea" . . . "Listen: a four-worded wavespeech: seesoo, hrss, rsseeiss, ooos." One of the most astonishing passages in all of literature is Joyce's playful answer to his own query: "What in water did Bloom, waterlover, drawer of water, watercarrier returning to the range, admire?"

The answer (in part):

Its universality: its democratic equality and constancy to its nature in seeking its own level: its vastness in the ocean of Mercator's projection: its unplumbed profundity in the Sundam trench of the Pacific exceeding 8,000 fathoms: the restlessness of its waves and surface particles visiting in turn all points of its seaboard: the independence of its units: the variability of states of sea: its hydrostatic quiescence in calm: its hydrokinetic turgidity in neap and spring tides: its subsidence after devastation: its sterility in the circumpolar icecaps, arctic and antarctic: its climatic and commercial significance: its preponderance of 3 to 1 over the dry land of the globe: its indisputable hegemony extending in square leagues over all the region below the subequatorial tropic of Capricorn; the multisecular stability of its primeval basin; its luteofulvous bed; its capacity to dissolve and hold in solution all soluble substances including millions of tons of the most precious metals: its slow erosions of peninsulas and downwardtrending promontories: its alluvial deposits: its weight and volume and density: its imperturbability in lagoons and highland tarns: its gradation of colours in the torrid and temperate and frigid zones: its vehicular ramifications in continental lakecontained streams and confluent oceanflowing rivers with their tributaries and transoceanic currents: gulfstream, north and south equatorial courses: its violence in seaquakes, waterspouts, artesian wells, eruptions, torrents, eddies, freshets, spates, groundswells, watersheds, waterpartings, geysers, cataracts, whirlpools, maelstroms, inundations, deluges, cloudbursts: its vast circumterrestrial ahorizontal curve . . . —(New York: Modern Library, 1961), pp. 671–72.

[p. 252] Kipling's letter revealing his horror of the Great Lakes is quoted also in Victoria Brehm's *Sweetwater, Storms, and Spirits: Stories of the Great Lakes,* p. 2.

Chapter 16

[p. 256] In spring 2002, the EPA announced a plan for solving some of the problems facing the Great Lakes. Entitled "Great Lakes Strategy 2002," the plan called for reducing concentrations of PCBs in lake trout and walleye by 25 percent within five years. It also proposed setting aside 300,000 acres of undeveloped land as buffer strips around farms and feedlots by the year 2007, thus reducing the amounts of solid wastes, fertilizers, and pesticides reaching the Great Lakes. In addition, the plan called for cleaning up public beaches (many of which currently suffer frequent closings due to solid waste contamination) by 2010, keeping 90 percent of the Great Lakes' six hundred swimming beaches open during at least 95 percent of the summers. Other strategies were proposed for preventing or at least slowing further invasions of non-native species. Unfortunately, "Great Lakes Strategy 2002" arrived without funding to implement its well-intentioned plan.

[p. 257] The statement about the depletion of oxygen in Lake Erie in the spring and summer of 2002—"This is the beginning of the second environmental war in Lake Erie"— is by David Rockwell, senior scientist at the U.S. EPA Great Lakes office, quoted in *The Toledo Blade,* June 18, 2002.

[pp. 261–262] The history of the sinking of the *Rising Sun* is from George Weeks's *Sleeping Bear: Yesterday and Today.*

Acknowledgments

I'm grateful to many people for their assistance during the research and writing of this book. I especially want to thank Rex Hite for the use of his foul-weather gear during the trip on the *Malabar* and for correcting technical errors in the manuscript. Thanks also to Dave Gerber for getting me aboard a sailboat in the Chicago-to-Mackinac Sailboat Race; to Dave Mull and Ron Barger for the days of fishing on Lake Erie; to Tom Carney for his frequent good counsel and his knowledge of Lake St. Clair and the St. Clair and Detroit rivers; to James McCullough for his excellent company in guiding me to various Hemingway haunts (and to a certain trout pond that shall remain unnamed); and to Glenn Wolff for the maps that grace the endpapers.

Thanks to Carol Swinehart of the Michigan Sea Grant College Program for answering my many questions and supplying me with dozens of publications; to Professor Ted Batterson at Michigan State University for sharing his lecture materials on the biology of the lakes; and to Dr. Michael Chiarappa of the Great Lakes Center for Maritime Studies for advice and background material on fisheries policy.

For their friendship, advice, and constant inspiration, I'm grateful to Glenn Wolff, Carole Simon, Kelly and Penny Galloup, Jim and Mary Ann

Linsell, Craig Date, Dave Scroppo, Norris McDowell, Bob Linsenman, Keith Taylor, Richard VanderVeen, Doug and Anne Stanton, and Carl and Eileen Ganter. For the support and encouragement only family can provide, I'm grateful to Gerald and Eva Dennis, Rick and Susan Dennis, Melissa and Eric Adams, and Arden and Virginia Johnson.

Finally, as always, for more reasons than I can list, my love and gratitude to Gail, Aaron, and Nick.

Index